AAOS

Jeff McDonald, NREMT-P

BLS Skills Review

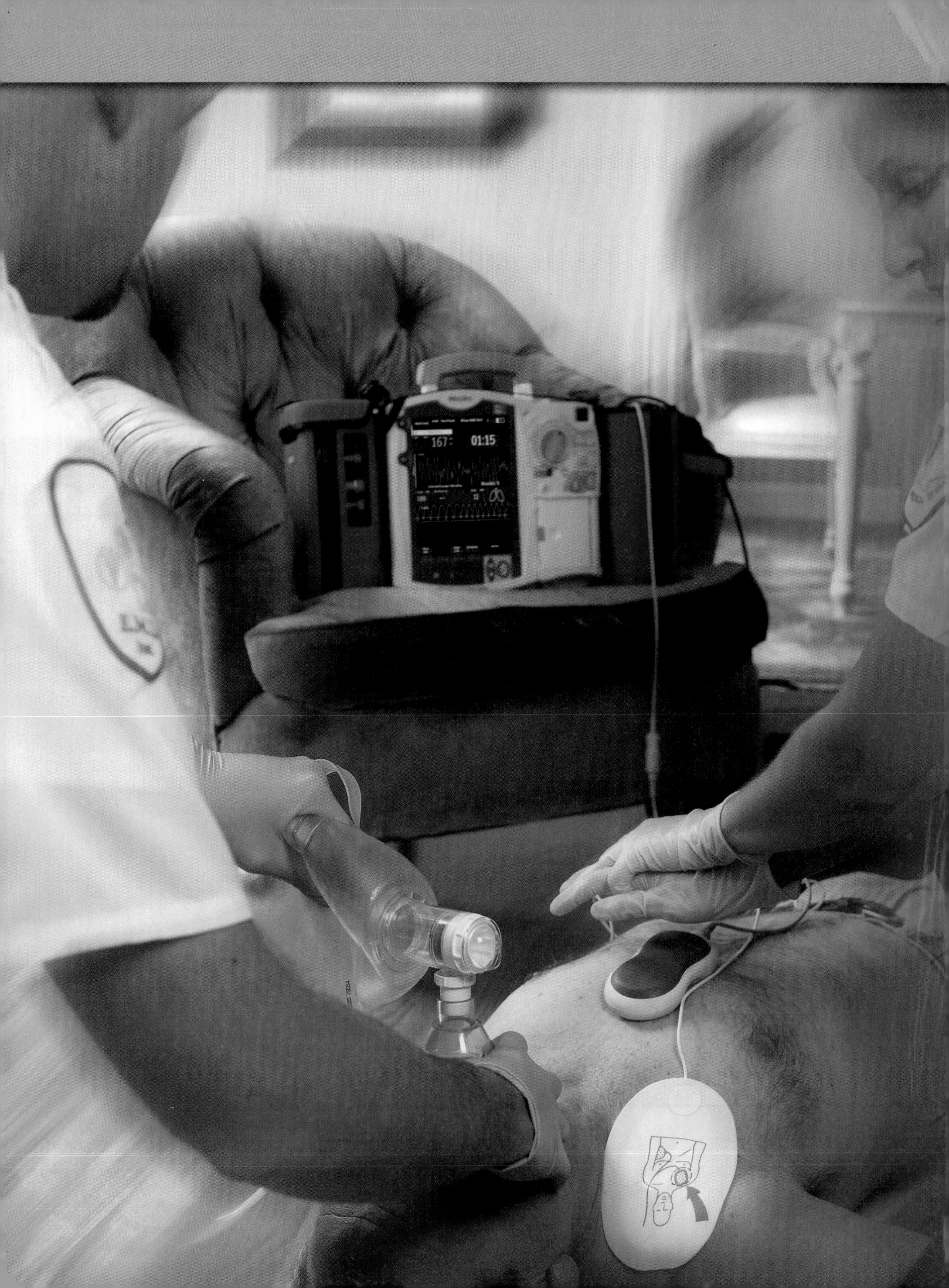
167
01:15

BLS Skills Review

AAOS
AMERICAN ACADEMY OF ORTHOPAEDIC SURGEONS

Jeff McDonald, NREMT-P
Author
Tarrant County College
Ft. Worth, Texas

Joseph A. Ciotola, MD, FAAOS
Medical Editor
Department of Emergency Services
Queen Anne's County, Maryland

JONES AND BARTLETT PUBLISHERS
Sudbury, Massachusetts
BOSTON TORONTO LONDON SINGAPORE

World Headquarters
Jones and Bartlett Publishers
40 Tall Pine Drive
Sudbury, MA 01776
978-443-5000
info@jbpub.com
www.jbpub.com

Jones and Bartlett Publishers Canada
6339 Ormindale Way
Mississauga, Ontario L5V 1J2
Canada

Jones and Bartlett Publishers International
Barb House, Barb Mews
London W6 7PA
United Kingdom

Jones and Bartlett's books and products are available through most bookstores and online booksellers. To contact Jones and Bartlett Publishers directly, call 800-832-0034, fax 978-443-8000, or visit our website, *www.jbpub.com*.
Substantial discounts on bulk quantities of Jones and Bartlett's publications are available to corporations, professional associations, and other qualified organizations. For details and specific discount information, contact the special sales department at Jones and Bartlett via the above contact information or send an email to specialsales@jbpub.com.

AAOS
AMERICAN ACADEMY OF ORTHOPAEDIC SURGEONS

Library of Congress Cataloging-in-Publication Data
McDonald, Jeff, 1960-
BLS skills review / Jeff McDonald.
p. ; cm.
"American Academy of Orthopaedic Surgeons."
ISBN 978-0-7637-4684-1 (pbk.)
1. CPR (First aid)--Outlines, syllabi, etc. 2. Emergency medicine--Outlines, syllabi, etc.
I. American Academy of Orthopaedic Surgeons. II. Title. III. Title: Basic life support skills review.
[DNLM: 1. Cardiopulmonary Resuscitation--Outlines. 2. Emergency Treatment--Outlines. WA 18.2 M479b 2008]
RC87.9.M37 2008
616.1'025--dc22

2007035336

6048
Printed in the United States of America
13 12 11 10 09 10 9 8 7 6 5 4 3 2

Contents

Acknowledgments

We would like to thank the following reviewers for their review of the *BLS Skills Review* manuscript:

Anthony J. Brunello, BS, RN, PHRN, TNS
Provena St. Mary's Hospital
Will/Grundy EMS System
Kankakee, Illinois

Raymond W. Burton, Retired
Plymouth Regional Police Academy
Plymouth, Massachusetts

Baruch S. Fertel, MPA, EMT-CIC
New York University
New York, New York

Joyce Foresman-Capuzzi, BSN, RN, CEN, CPN, CTRN, EMT-P
Temple University Health System Transport Team
Philadelphia, Pennsylvania

Gordon J. Elquist, AS, L-EMT-P
El Paso Fire Department
Airport Training Center
El Paso, Texas

Donell Harvin, MPA, MPH, EMT-P
New York City Department of Health
New York, New York

Les Hawthorne, BA, NREMT-P
Southwestern Illinois College
Belleville, Illinois

K. C. Jones, BSE, NREMT-P
North Arkansas College
Harrison, Arkansas

Alan Joos, EMT-I, EFO, MS
Columbia Southern University
Orange Beach, Alabama

Angela Kruep, EMT-B
Pima Community College
Tucson, Arizona

John A. Kubincanek, EMT-P, AA, EMSI
Cuyahoga Community College
Cleveland, Ohio

Donald G. Lee, Jr., BS, CP
Pensacola Junior College
Pensacola, Florida

David J. Leven, NYS EMT-B
University of Rochester
Rochester, New York

J. J. Magyar, MS, NREMT-P
Pennsylvania College of Technology
Williamsport, Pennsylvania

Mike McEvoy, PhD, EMT-P, RN, CCRN
Saratoga County EMS
Saratoga, New York

Beth Ann McNeill, BFA, EMT-B, NYS CIC
Monroe Community College
Rochester, New York

Robert J. Staples, MS, NREMT-P
Seminole Community College
Sanford, Florida

Photo Credits

Student Resources

BLS SKILLS REVIEW DVD

ISBN 13: 978-0-7637-5223-1

Designed for individual student use, this action-packed DVD demonstrates the proper techniques for each skill presented in the manual. Capturing real-life scenes, each skill is clearly broken down, demonstrated, and applied in a variety of drills. This DVD is an invaluable resource for every BLS provider and gives students a chance to witness providers in action and in "real" time.

Note to Reader:

The skills depicted in this book reflect a training situation. Other or additional safety precautions may be required in an actual emergency.

Instructor Resources

INSTRUCTOR'S TEACHING PACKAGE
ISBN 13: 978-0-7637-5700-7

Created for instructors to use in skills labs, the teaching package contains a copy of the companion DVD and an Instructor's ToolKit CD-ROM with:

- **Customizable PowerPoint Presentations** These slides follow the skills presented in the manual and include images to enhance the student's classroom experience. Slides can be modified and edited to meet your needs.
- **Comprehensive Image Bank** Use these images to further enhance your PowerPoint presentations, make handouts, or highlight skills for further classroom discussion.
- **Skill Evaluation Sheets** These sheets allow you to track your students' performance. They also make an ideal study tool.

The measure of our greatness is not in saving those lives that everyone else can save, for that is the standard of care.

Our greatness is measured by saving those that no one else can save.

—Roy K. Yamada, MD

Personal Safety

Personal Protective Equipment

Introduction

Personal Protective Equipment (PPE) is used to provide a barrier between the EMS responder and infectious agents that may be found in blood, body fluids, and exhaled air. It is the responsibility of the employer to provide EMS personnel with personal protective equipment (gloves, eye protection, masks, and gowns) sufficient to provide reasonable protection. It is the responsibility of EMS personnel to wear the personal protective equipment at the appropriate times and in the appropriate manner.

It is important for EMS personnel to understand that we work in an uncontrolled environment. To provide the best protection for unseen dangers, EMS personnel should enter every call with the assumption that the patient has an infectious disease and have appropriate PPE in place at the time of patient contact.

Procedures

Step 1 Prepare for patient contact.

Prior to making patient contact, don protective gloves and eye protection. This should be performed on every patient regardless of risk of infectious disease.

Because of the prevalence of latex allergies, it is best to always use nonlatex gloves. This will prevent the development of latex allergies in the wearer and the stimulation of a reaction in latex-sensitive patients.

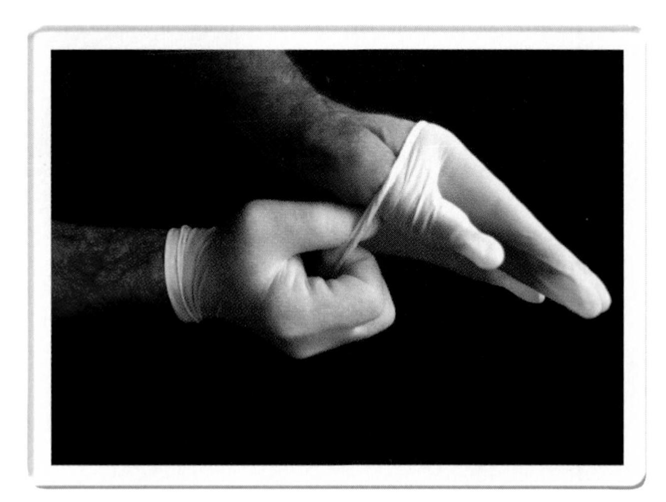

Getting into the habit of always wearing eye protection is a must. This will protect your eyes from unforeseen infectious exposures.

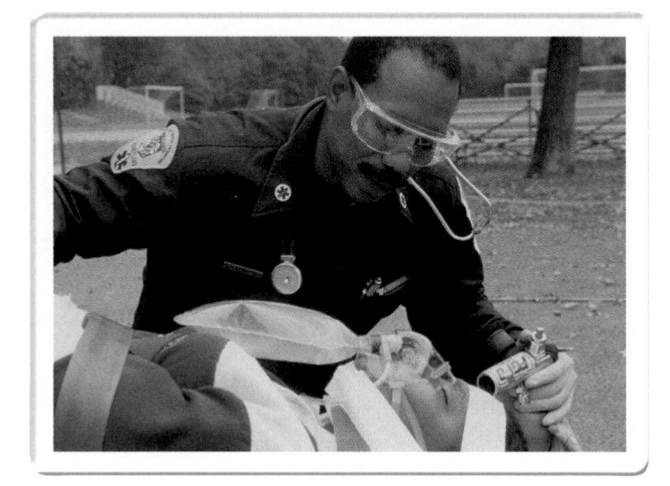

Safety Tips

Handwashing and Gloves

Studies have shown that gloves provide a barrier, but that neither vinyl nor latex procedure gloves are completely impermeable. It is important that you wash your hands every time you remove a pair of gloves.

Step 2 Add personal protective equipment as needed.

Upon initial contact, assess the need for additional body substance isolation. Don as appropriate the following pieces of equipment:

- **Masks or barrier devices** should be used for any invasive procedure requiring close facial contact. For instance, a pocket mask should be used to provide artificial ventilations.

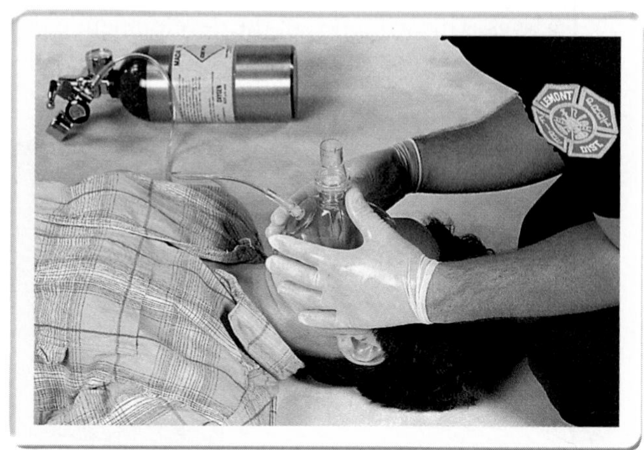

- **HEPA filter masks** should be worn any time your patient has or is suspected to have tuberculosis.

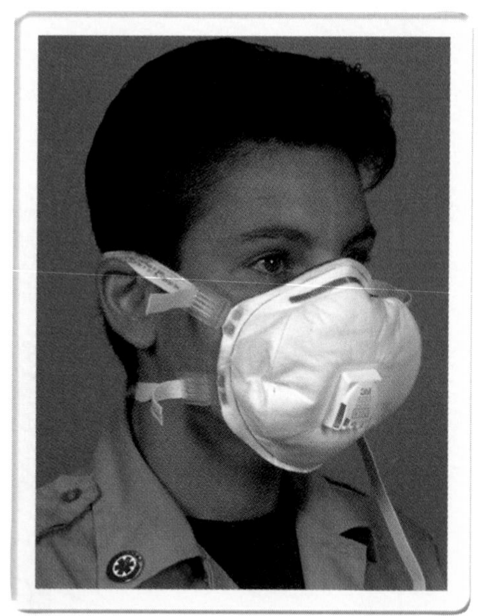

- **Gowns** should be worn to protect duty uniforms from becoming contaminated by blood or body fluids.

Body Substance Isolation

Introduction

Body Substance Isolation (BSI) is the process of containing blood and other body fluids that may harbor infectious agents. The process starts by wearing personal protective equipment, but also includes procedures for containment, cleanup, and disposal of contaminated supplies and other objects.

Procedures

Step 1 Remove contaminated clothing.

Gowns and other protective devices prevent the contamination of clothing. However, in cases where clothing does become contaminated it is important to remove the clothing carefully. Shirts should never be removed over the head. Pullover shirts may need to be cut off and destroyed.

Once contaminated clothing is removed it should be bagged as a contaminated material. The clothing must be properly laundered and decontaminated by the EMS employer and not by the employee's home laundry.

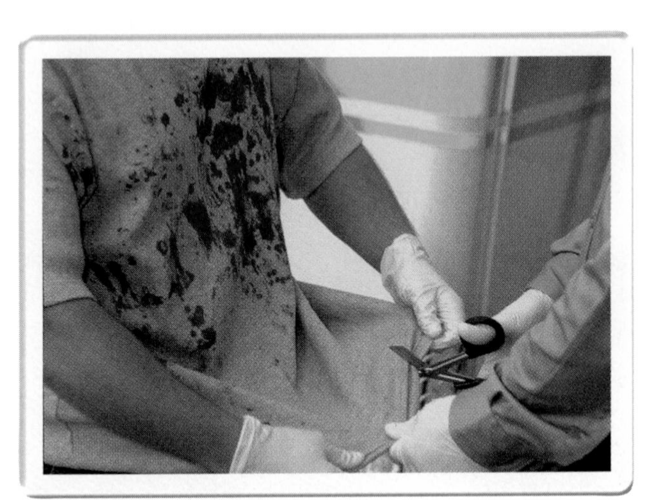

Step 2 Clean spilled blood and body fluids.

Spilled blood and body fluids must be cleaned and the area disinfected before the equipment is returned to service. Depending on the size of the spill, it may be as simple as applying a bleach solution and a paper towel or as complicated as applying gallons of absorbent gels. In either case, it is important to place all materials used in the cleanup in biohazard bags for proper disposal.

While cleaning up body fluid spills, it is important to wear gloves and eye protection. Standard EMS gloves are sufficient for light spills with simple disinfectants. Larger spills where strong disinfectants are used may require thicker gloves intended for use with harsh chemicals.

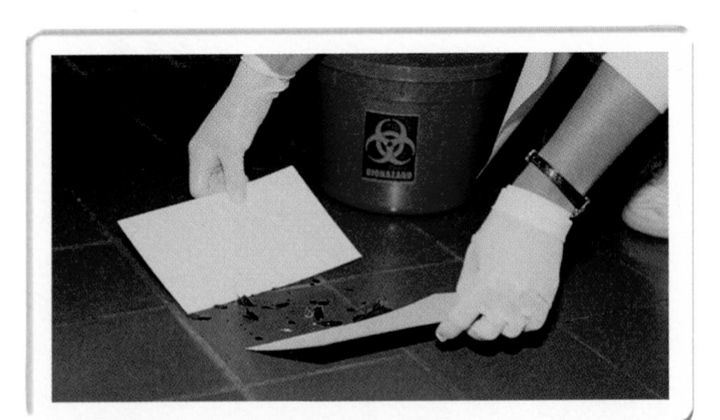

Handwashing

Introduction

Handwashing is needed as a component in the preparation for, or the cleanup following patient contact.

Procedures

Grasp towel.

Remove towel from towel dispenser.

Turn on water, discard towel.

Use towel to turn water on. Discard towel into waste bin.

Apply soap and scrub vigorously.

Apply soap to both hands and bring to lather. This may be done before or after wetting the hands depending on the type of soap used. Scrub vigorously for at least 15 seconds. When preparing for sterile procedures or preparing for delivery the hands should be scrubbed for at least 2 minutes, aided by brush or sponge.

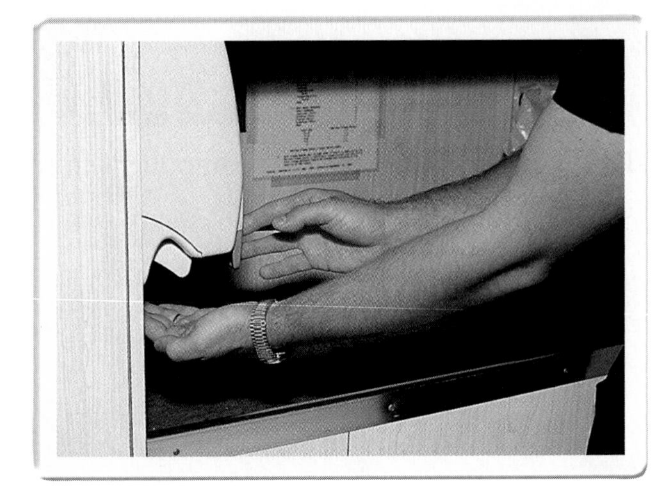

Rinse thoroughly.

Rinse hands well.

Grasp towel and dry.

Remove clean towel from towel dispenser. Dry your hands well. Several towels may be needed.

Step 6 Dispose of towel.

Discard towel into waste bin.

Step 7 Grasp towel and turn off water.

Remove clean towel from towel dispenser and turn water off. Discard towel into waste bin.

Step 8 Grasp towel and open door.

Remove clean towel from towel dispenser and open the door.

Step 9 Dispose of towel.

Discard towel into waste bin.

In the Field

Waterless Cleansers

It is often difficult to find water in the prehospital environment. In these cases the use of waterless, alcohol-based cleansers can be very beneficial. They should be used routinely when the driver of the ambulance removes his or her gloves and enters the cab of the ambulance.

Airway Management

Introduction

Airway, oxygenation, and ventilation management, after our own personal safety, are the most important skills in patient care.

Airway management means employing airway and ventilation skills appropriate to the current needs of the patient, and often in conjunction with other skills. In trauma management we have seen a shift from thinking that cervical spine immobilization precedes the management of the airway. In actuality, it is important to use these skills together, managing the airway with concern for the cervical spine.

As important as airway is, oxygenation and ventilation are equally as important. In fact, airway accomplishes nothing if oxygenation and ventilation are not included. Oxygenation and ventilation must always be used at proper rates and adjusted to meet the needs of the individual patient.

Airway, oxygenation, and ventilation are the primary patient care responsibility of emergency medical services personnel. The skills in this section will help you to use airway management skills appropriately and in a timely manner.

SECTION 1

SKILL 1

Jaw-Thrust Maneuver

Performance Objective

Given an adult patient, the candidate shall demonstrate the use of the jaw-thrust maneuver using the criteria herein prescribed, in 1 minute or less.

Equipment

The following equipment is required to perform this skill:

- Appropriate body substance isolation/personal protective equipment

Equipment that may be helpful:

- Oropharyngeal airways (various sizes)
- Nasopharyngeal airways (various sizes)
- Bag-mask device
- Oxygen cylinder
- Oxygen regulator

Indications

- Immediate opening of the airway in unresponsive patients prior to determining cause of unconsciousness and with potential for spine injury
- Maintenance of the airway in nonbreathing trauma patients

Contraindications

- Difficulty achieving an open airway using the procedure. The International Liason Committee on Resuscitation (ILCOR) recommends that if you cannot achieve an open airway with the jaw-thrust maneuver, you should abandon the procedure and use the head tilt–chin lift maneuver (see Skill 3).

Complications

- None if properly performed and maintained

Procedures

Step 1 Ensure body substance isolation before beginning procedures.

Prior to beginning patient care, appropriate body substance isolation procedures should be employed.

Step 2 Assume proper position.

Position yourself at the patient's head, facing the long axis of the body.

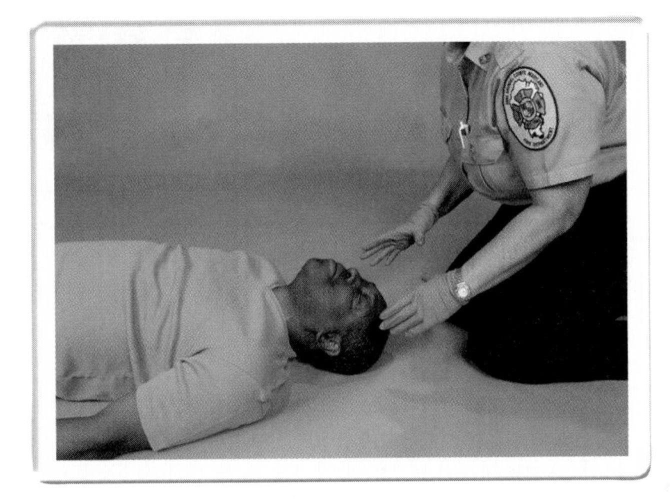

Step 3 Position hands on face.

Using both hands, place your thumbs on the cheek bones with the base of the thumb just below the patient's eyes. Reach forward with your index fingers and place them behind the lower jaw just below the ear.

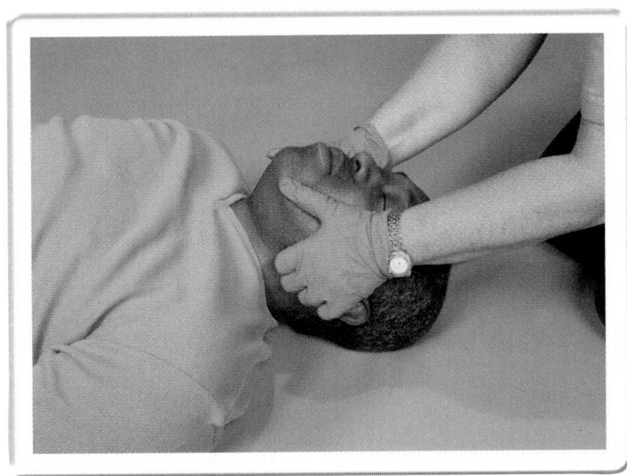

Thrust jaw.

Without moving the patient's neck, push the lower jaw forward (anteriorly) to lift the tongue off the posterior pharynx. It is important to maintain this position until airway adjuncts maintain the airway.

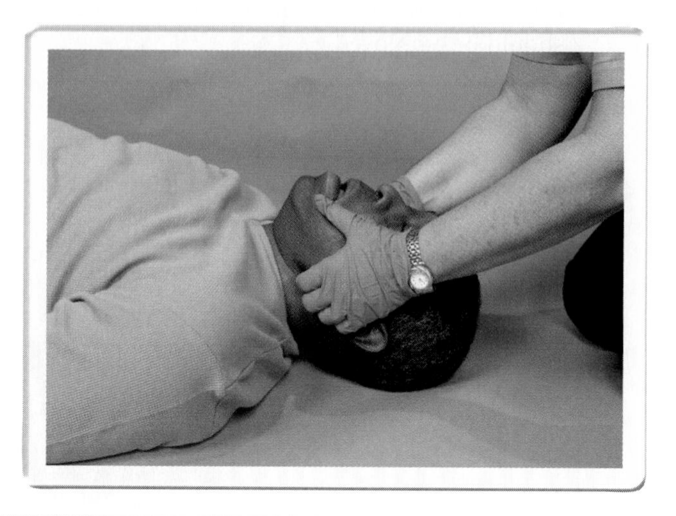

Assess breathing.

Once the airway is open, assess the patient's ability to breathe spontaneously. If the patient is breathing on his or her own, maintain the open airway until artificial adjuncts can be inserted. If the patient is not breathing, begin artificial ventilation using either mouth-to-mask or bag-mask ventilation.

Tongue–Jaw Lift Maneuver

SKILL 2

Performance Objective

Given an adult patient, the candidate shall demonstrate the use of the tongue–jaw lift maneuver using the criteria herein prescribed, in 1 minute or less.

Equipment

The following equipment is required to perform this skill:

- Appropriate body substance isolation/personal protective equipment

Equipment that may be helpful:

- Oropharyngeal airways (various sizes)
- Nasopharyngeal airways (various sizes)
- Bag-mask device
- Oxygen cylinder
- Oxygen regulator

Indications

- Immediate opening of the airway in unresponsive patients prior to determining cause of unconsciousness
- Airway obstruction caused by the tongue in supine, unresponsive patients

Contraindications

- Nonbreathing patients

Complications

- None if properly performed and maintained

Procedures

Step 1 Ensure body substance isolation before beginning procedures.

Prior to beginning patient care, appropriate body substance isolation procedures should be employed.

Step 2 Assume proper position.

Position yourself at the patient's head and shoulder, facing across the body.

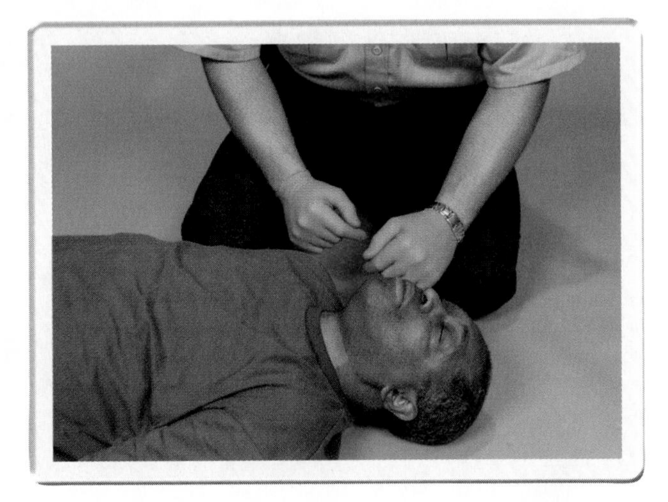

To ensure that the cervical spine does not move, place the hand closest to the patient's head on the patient's forehead. Confirm that the patient is unresponsive before beginning the next step of the procedure.

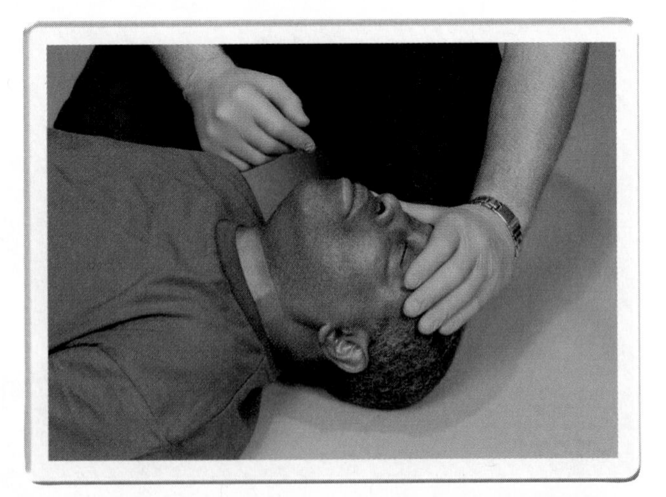

Step 3 Grasp tongue and jaw.

With the hand closest to the patient's feet, insert your thumb into the patient's mouth.

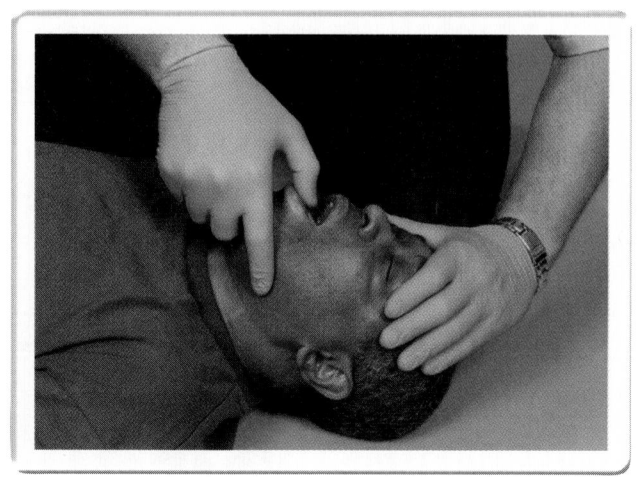

Grasp the tongue under the thumb and keep the index finger tight against the underside of the lower jaw.

Step 4 Lift tongue and jaw.

Gently lift the tongue and jaw to pull the tongue straight up from the posterior wall of the pharynx.

Step 5 Assess breathing.

Once the airway is open, assess the patient's ability to breathe spontaneously. If the patient is breathing, use the jaw-thrust maneuver to maintain the airway until artificial adjuncts can maintain the airway. If the patient is not breathing switch to the jaw-thrust maneuver and begin artificial ventilation using mouth-to-mask or bag-mask ventilation (see Skill 6).

Safety Tips

Grasping and Lifting the Tongue and Jaw

Injury may occur to the rescuer's thumb or finger if the patient begins biting motions or has seizure activity. Use extreme caution when using this procedure.

SKILL 3

Head Tilt–Chin Lift Maneuver

Performance Objective

Given an adult patient, the candidate shall demonstrate the use of the head tilt-chin lift maneuver using the criteria herein prescribed, in 1 minute or less.

Equipment

The following equipment is required to perform this skill:

- Appropriate body substance isolation/personal protective equipment

Equipment that may be helpful:

- Oropharyngeal airways (various sizes)
- Nasopharyngeal airways (various sizes)
- Bag-mask device
- Oxygen cylinder
- Oxygen regulator

Indications

- Immediate opening of the airway in unresponsive patients with no potential for cervical spine injury

Contraindications

- Patients with a potential or known cervical spine injury

Complications

- None if properly performed and maintained

Procedures

 Step 1

Ensure body substance isolation before beginning procedures.

Prior to beginning patient care, appropriate body substance isolation procedures should be employed.

  Step 2

Assume proper position.

Position yourself at the patient's head and shoulder, facing across the body.

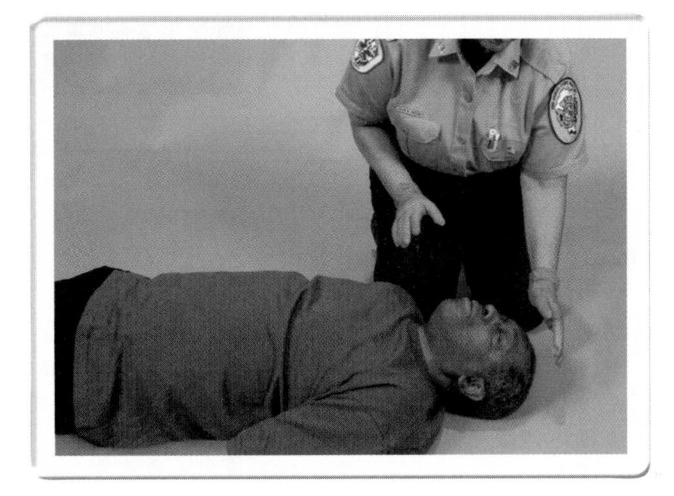

Place the hand closest to the patient's head on the patient's forehead.

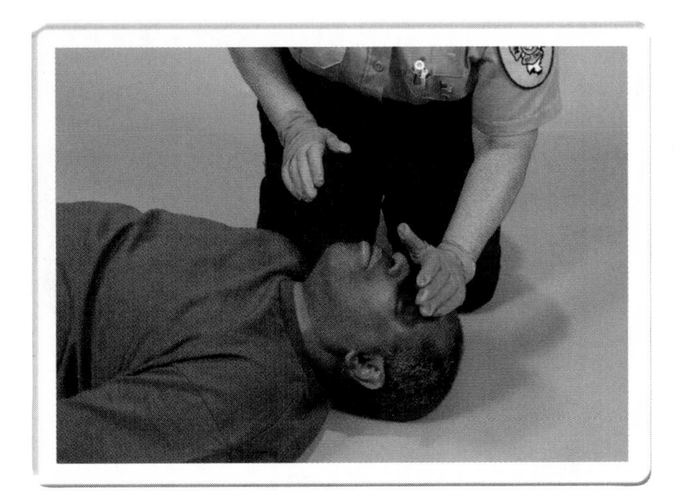

Step 2 continued

Place the hand closest to the patient's feet on the underside of the mandible just below the chin.

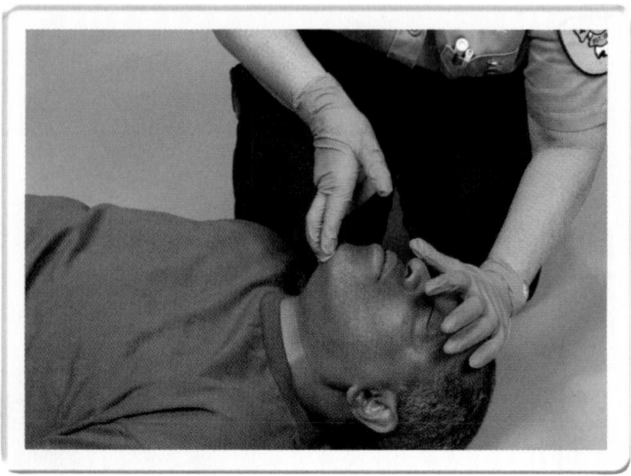

Step 3 Hyperextend the neck.

Maintaining hand position, pull up on the patient's chin while pushing down on the patient's forehead. Hyperextend the neck until the face is at least 45° to the floor. It is important to maintain this position until airway adjuncts maintain the airway.

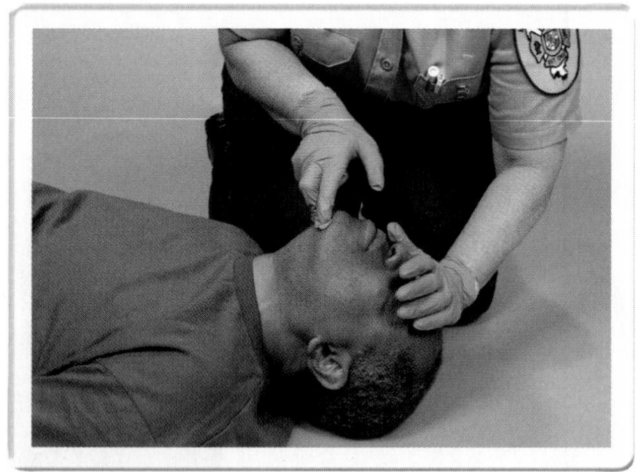

Step 4 Assess breathing.

Once the airway is open, assess the patient's ability to breathe spontaneously. If the patient is breathing, maintain the airway until artificial adjuncts can maintain the airway. If the patient is not breathing, begin artificial ventilation using mouth-to-mask or bag-mask ventilation.

Oxygen Administration

Performance Objective

Given an adult patient and appropriate oxygen delivery devices, the candidate shall demonstrate proper attachment of the regulator, ability to read the pressure gauge accurately, and ability to deliver oxygen using the criteria herein prescribed, in 5 minutes or less.

Equipment

The following equipment is required to perform this skill:

- Appropriate body substance isolation/personal protective equipment
- Oxygen cylinder, regulator, and key
- Oxygen delivery device (appropriate to patient)
  - Nasal cannula
  - Simple face mask
  - Nonrebreathing mask
  - Bag-mask device

Equipment that may be helpful:

- Pulse oximeter
- End-tidal carbon dioxide meter

Indications

- Oxygen supplement for patients in:
  - Respiratory distress
  - Respiratory failure
  - Respiratory arrest
  - General hypoxia
  - Shock
  - Myocardial infarction
  - Stroke

Contraindications

- Paraquat poisoning

Complications

- Rate dependant. High-flow oxygen levels can cause constriction of cerebral and coronary vessels. It is important to base oxygen administration on assessment of arterial oxyhemoglobin saturation (Spo_2) and end-tidal carbon dioxide concentration.

Safety Tips

Oxygen and Petroleum: A Dangerous Mix

Oxygen and petroleum don't mix and can be a very dangerous combination. Never lubricate oxygen regulators, never apply lubricant to oxygen O-rings, and never administer oxygen with oily hands. Wipe petroleum jelly or lip balm from patients' lips before placing oxygen masks.

Safety Tips

Understanding Oxygen as a Medication

Each drug carried on your EMS unit has specific uses. It is *essential* for all EMS personnel who carry drugs on their unit to understand all the indications, contraindications, side effects, precautions, doses, and routes of administration for each drug carried. This includes oxygen. We often forget that oxygen is a drug and as such has specific indications, contraindications, cautions, and side effects. Old-school approaches to oxygen administration led us to believe that "if a little oxygen was good, more is better." Nothing could be further from the truth. The dosing of oxygen must be tailored and adjusted to the situation. Although any patient experiencing dyspnea may benefit from the administration of oxygen, it is often ventilation, not oxygen, that is more important. Blindly giving oxygen, especially high levels of oxygen, may be detrimental to the patient.

Procedures

Step 1 Ensure body substance isolation before beginning procedures.

Prior to beginning patient care, appropriate body substance isolation procedures should be employed.

Step 2 Assemble regulator to tank.

It is important to be able to recognize the cylinder as an oxygen cylinder by color (green, white, or chrome) and pin index (2:5), and to ensure that the cylinder is labeled as medical oxygen (Oxygen USP).

Point the oxygen port away from self, patient, and bystanders so no debris gets blown at a person (or pet). Open and close the valve quickly to blow out debris.

Confirm the presence of an O-ring at the oxygen port.

Procedures

Step 1 Ensure body substance isolation before beginning procedures.

Prior to beginning patient care, appropriate body substance isolation procedures should be employed.

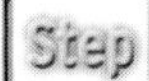 

Step 2 Connect one-way valve to mask.

Assemble the mask by forming the cup and connecting the one-way valve.

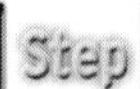 

Step 3 Open patient's airway or confirm patient's airway is open (manually or with adjunct).

Before opening the airway, consider the possibility of cervical spine injury. If spinal injury is suspected, use a jaw-thrust maneuver to open the airway (see Skill 1). If no spinal cord injury is suspected, open the patient's airway using the head tilt–chin lift maneuver (see Skill 3).

An oropharyngeal airway may be employed if available. Although not necessary initially, long-term ventilations should include its use.

Step 4 Establish and maintain a proper mask-to-face seal.

Position the mask with the apex over the bridge of the nose and the base between the lower lip and the prominence of the chin. Using firm pressure around the sides of the mask, form a tight seal to prevent air leakage.

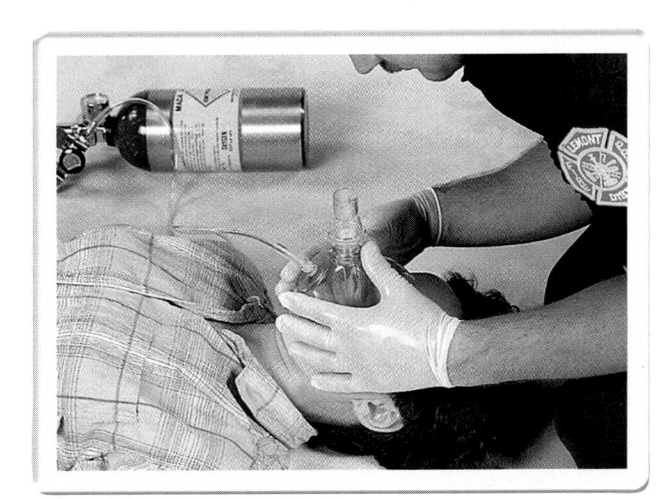

Step 5 Ventilate the patient at the proper volume and rate.

Begin ventilations as soon as the mask is sealed and assess for air leakage. Effective ventilations (ie, those breaths that cause the chest to rise adequately for the size of the patient) must be started within 30 seconds. Ventilate at a rate of 10 to 12 breaths/min.

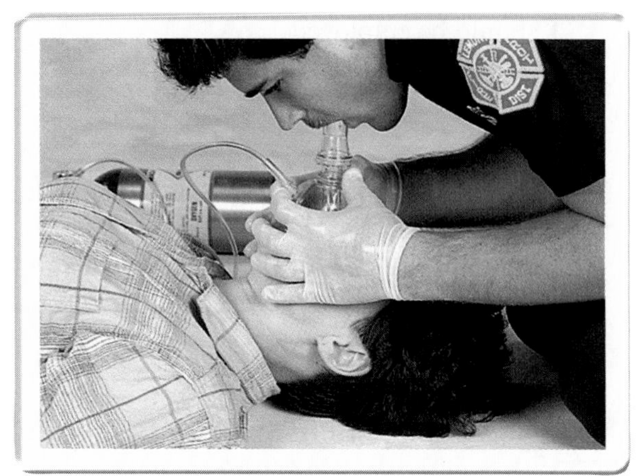

Watch for chest rise and fall during exhalation.

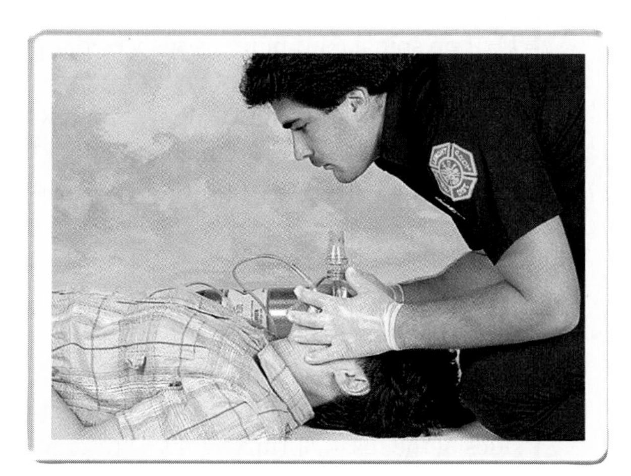

Step 6 Connect mask to high-concentration oxygen.

Attach oxygen tubing to inlet on mask and oxygen source.

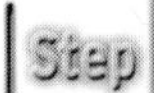 

Step 7 Adjust flow rate to 15 L/min.

Open the cylinder valve and adjust liter flow to 15 L/min.

Step 8 Ventilate the patient at the proper volume and rate.

Resume effective ventilations within 15 seconds of last breath. Be careful not to overventilate. Each ventilation should be delivered slowly and easily, lasting 1 full second. Fast and high-pressure ventilations can cause air to enter the stomach, increasing the risk of vomiting. If a second rescuer is available, the Sellick maneuver (cricoid pressure) may be applied.

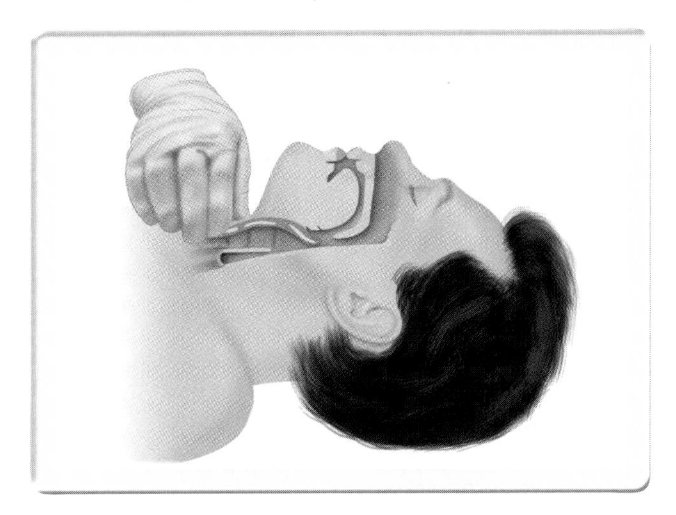

Continue the ventilations at a rate of 10 to 12 breaths/min until the patient begins spontaneous respirations, you are exhausted, you are relieved, or you are told to stop.

SKILL 6

Bag-Mask Ventilation

Performance Objective

Given an adult patient, a bag-mask device, and all appropriate equipment, the candidate shall demonstrate artificial ventilation using the criteria herein prescribed, in 5 minutes or less, with and without assistance from additional rescuers.

Equipment

The following equipment is required to perform this skill:

- Appropriate body substance isolation/personal protective equipment
- Oxygen cylinder
- Oxygen regulator
- Oxygen key
- Oxygen connecting tubing
- Bag-mask device with appropriate-sized mask
- Oropharyngeal airway (various sizes)

Equipment that may be helpful:

- Suction device
- Pulse oximeter
- End-tidal carbon dioxide meter

Indications

- Assisting the patient with shallow or slow respirations
- Providing ventilations to the nonbreathing patient

Contraindications

- None when properly applied

Complications

- May cause abdominal distention and subsequent vomiting if proper airway position is not maintained

Procedures

Step 1 Ensure body substance isolation before beginning procedures.

Prior to beginning patient care, appropriate body substance isolation procedures should be employed.

Step 2 Open the airway.

Before opening the airway, consider the possibility of cervical spine injury. If spinal injury is suspected, use a jaw-thrust maneuver to open the airway.

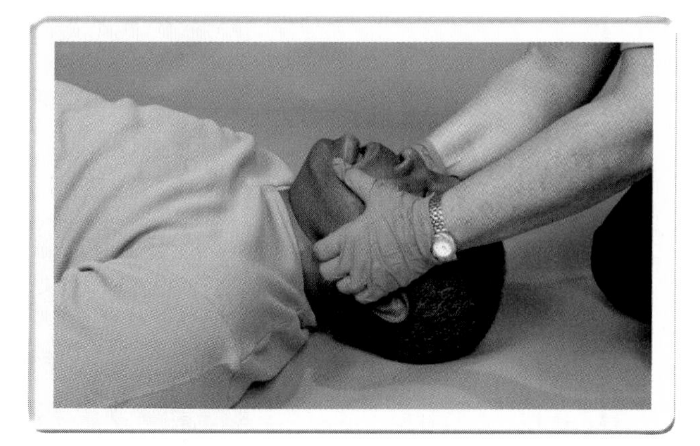

If no spinal cord injury is suspected, open the patient's airway using the head tilt–chin lift maneuver.

Step 3 Insert appropriate-sized oropharyngeal airway.

Measuring from the corner of the mouth to the base of the earlobe, choose the correct-sized airway for your patient.

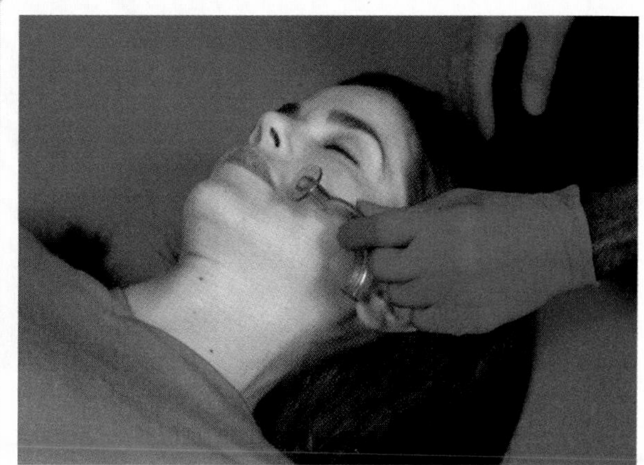

continued

Insert airway in front of mouth with tip pointed toward roof of mouth, or insert from side of mouth with tip toward inside of cheek.

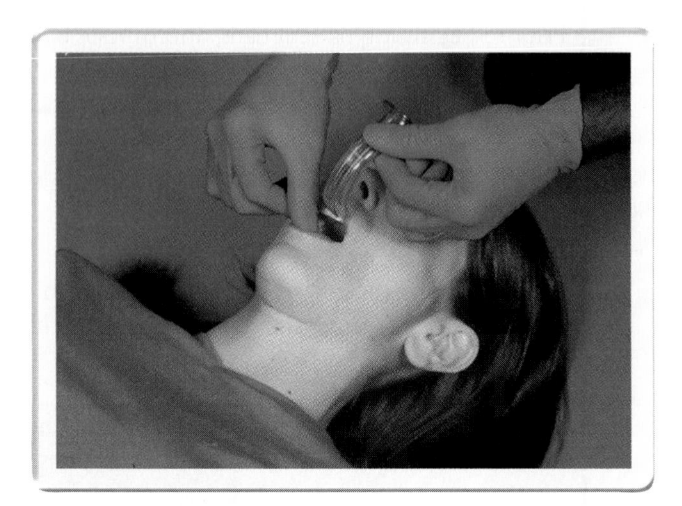

Rotate the tip of the airway toward the patient's feet as it is advanced into the oropharynx. When completely inserted, the flange of the airway should rest on the patient's lips.

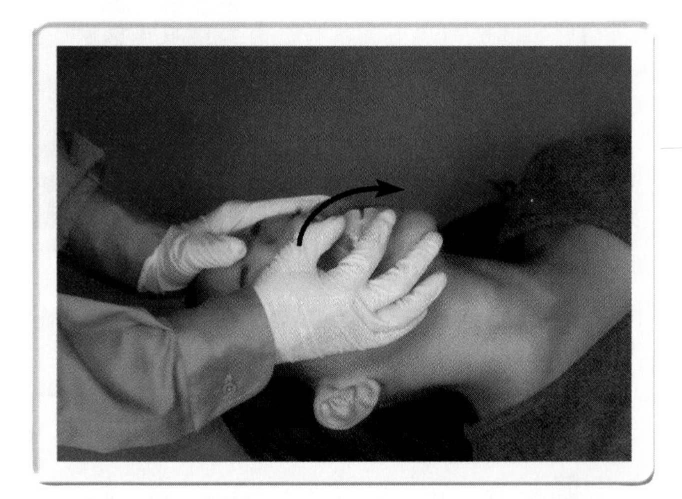

A tongue depressor can be used to hold the tongue inferior and the airway inserted tip down. If the patient starts to gag, remove the airway by pulling the flange anterior and inferior.

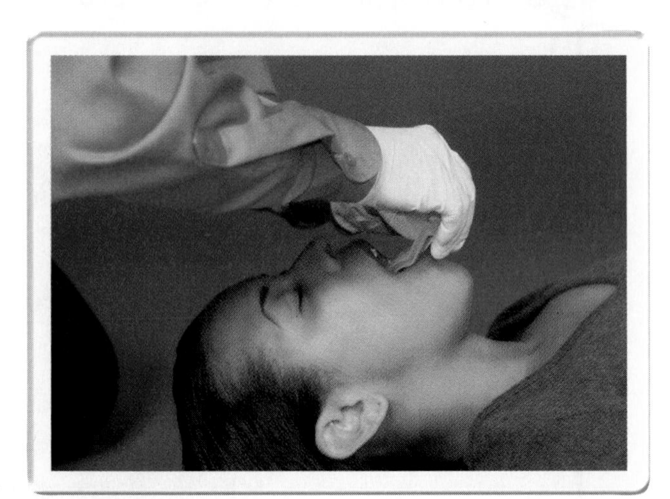

Step 4 Select appropriate-sized mask and bag.

Select the appropriate-sized bag for the patient. Bag and mask sizes are normally listed as adult, pediatric, and infant. These sizes are adequate for the average-sized person meeting the age definition criteria. However, large children will require a bag that will deliver the appropriate volume. Likewise, pediatric bags may be appropriate for small adults or adults in whom overpressurization could cause pulmonary damage. Masks should be chosen to fit the patient without air leakage.

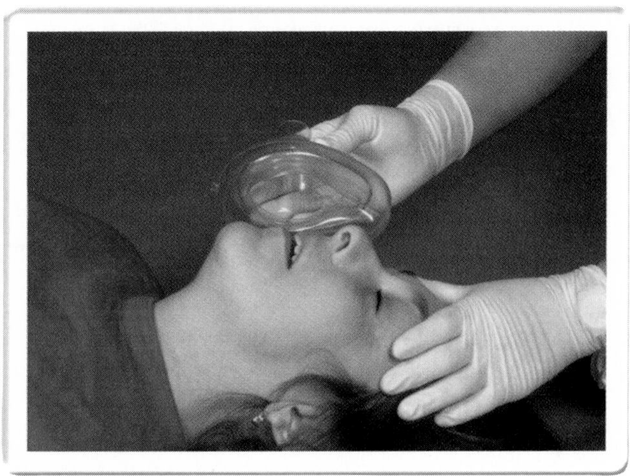

Step 5 Create a proper mask-to-face seal.

Position the mask with the apex over the bridge of the nose and the base between the lower lip and the prominence of the chin. Although this procedure can be performed by a single person, bag-mask ventilation should be performed by two people whenever possible. The first rescuer will assume the responsibilities of obtaining and maintaining the mask-to-face seal.

In the Field

Ventilating the Patient

Once the patient is intubated, ventilations can be performed by a single rescuer. Ensure that ventilations are slow and meticulous, adjusting the rate based on the patient's end-tidal carbon dioxide concentration if available and allowed under local protocol.

Step 6 Ventilate patient at no less than 800 mL volume, and between 10 and 20 breaths/min.

Begin ventilations as soon as the mask is sealed and assess for air leakage. For adults, deliver ventilations at a rate of 10 to 20 breaths/min with a tidal volume of at least 800 mL. Pay close attention to ensure that each ventilation is of the appropriate volume and of a consistent rate. Allow adequate exhalation between each breath.

If a second rescuer is available, he or she should assume the responsibilities of appropriate ventilations.

Effective ventilations (ie, those breaths that cause the chest to rise adequately for the size of the patient) must be started within 30 seconds. Be careful not to overventilate. Each ventilation should be delivered slowly and easily, lasting 1 full second. Fast and high-pressure ventilations can cause air to enter the stomach, increasing the risk of vomiting.

Step 7 Connect reservoir and oxygen.

Attach oxygen tubing to inlet on bag-mask device and to oxygen source.

Special Populations

Ventilating the Patient With a Tracheostomy

You must make a few simple adjustments to ventilate the patient with a stoma or tracheostomy tube. For patients with a tracheostomy tube, the bag-mask device can be attached directly and ventilated as you would an endotracheal tube. Tracheostomy tubes require specialized care and can easily become clogged with mucus. If you experience difficulty ventilating the patient, you may need to suction the tracheostomy tube using a flexible suction catheter. For patients without a tracheostomy tube, place a pediatric mask over the stoma on the patient's neck. Be careful to use sufficient pressure to achieve a good seal, but not so much pressure that you occlude circulation through the carotid arteries or jugular veins.

Step 8 Adjust liter flow to 15 L/min.

Open the cylinder valve and adjust liter flow to 15 L/min.

Step 9 Reopen the airway.

Using the same procedure as seen in Step 2 (keeping cervical spine injury in mind if suspected), reopen the patient's airway.

Step 10 Create a proper mask-to-face seal.

Position the mask with the apex over the bridge of the nose and the base between the lower lip and the prominence of the chin.

Step 11 Instruct assistant to begin ventilations.

If a second rescuer is available, he or she should ventilate the patient while the first rescuer maintains the airway and mask seal (see Step 5). Two rescuers should always be used when available because this provides better control of the mask seal and of the volume delivered. Both rescuers are responsible for monitoring proper rate and depth of each ventilation. If a third rescuer is available, the Sellick maneuver (cricoid pressure) should be applied.

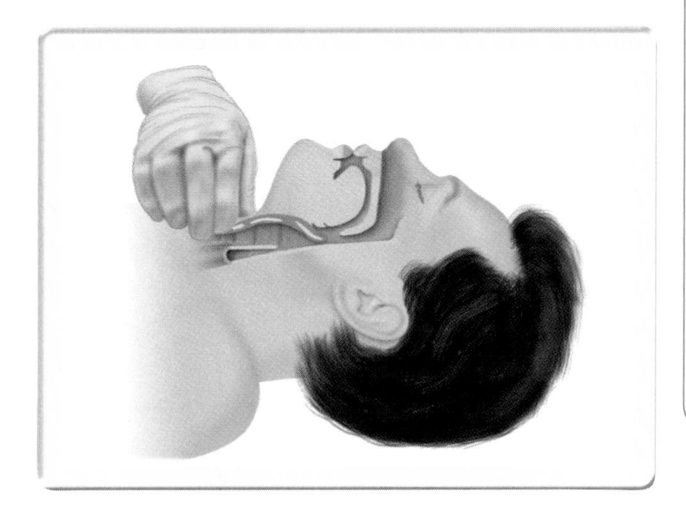

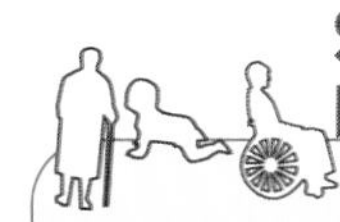

Special Populations

Ventilating Geriatric Patients

When ventilating the geriatric patient with known or suspected lung disease, consider using a pediatric bag-mask device instead of the adult bag. The use of a pediatric bag will prevent the patient from receiving too deep a ventilation, which could cause injury to weakened lungs. Confirm that this will be acceptable by local protocol and that the patient's ventilations are sufficient. Monitoring of the patient's condition, arterial oxyhemoglobin saturation (Spo_2), and end-tidal carbon dioxide concentration is important.

In the Field

Determining Appropriate Tidal Volume

The best method of determining adequate ventilations is to watch for effective, not excessive, chest rise. Since most bag-mask devices have no direct means of identifying how many milliliters are being delivered, direct calculation is not possible. However, knowing how much should be delivered and estimating the size and abilities of your bag, you can better ensure that adequate volumes are being delivered. The average person uses between 7 to 10 mL/kg for every breath. Calculate the tidal volume by multiplying the patient's weight in kilograms by 7. This will provide an effective volume for most patients. If chest rise seems minimal, increase the depth slightly.

SKILL 7

Oropharyngeal Airway

Performance Objective

Given an adult patient and appropriate equipment, the candidate shall demonstrate proper procedures for placing an oropharyngeal airway using the criteria herein prescribed, in 1 minute or less.

Equipment

The following equipment is required to perform this skill:

- Appropriate body substance isolation/personal protective equipment
- Oropharyngeal airway (various sizes)

Equipment that may be helpful:

- Tongue depressor
- Bag-mask device
- Oxygen cylinder
- Oxygen regulator

Indications

- Basic airway adjunct used to keep the tongue from occluding the airway in unresponsive patients
- Should be used in all unresponsive patients being ventilated with a bag-mask device

Contraindications

- Patients with an intact gag reflex
- Patients with foreign body airway obstruction

Complications

- If not properly inserted, may cause the tongue to be pushed back into the hypopharynx
- Aspiration may occur if the airway is not removed before the patient regains consciousness

Procedures

Step 1 Ensure body substance isolation before beginning procedures.

Prior to beginning patient care, appropriate body substance isolation procedures should be employed.

Step 2 Select appropriate-sized airway.

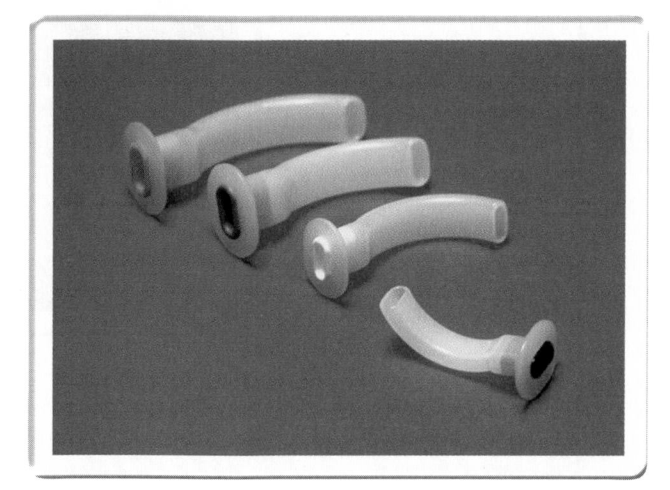

Measuring from the corner of the mouth to the base of the earlobe or angle of the jaw, choose the correct-sized airway for your patient.

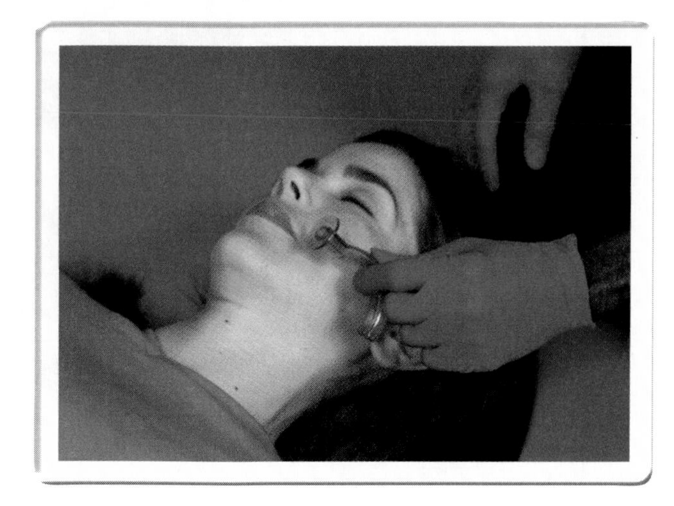

Safely insert airway without pushing tongue posteriorly (must create patent airway).

Before opening the airway, consider the possibility of cervical spine injury. If spinal injury is suspected, use caution opening the airway and maintain inline cervical stabilization. Open the mouth using the jaw-thrust or cross-finger technique. Insert the airway with the tip pointed toward the hard palate.

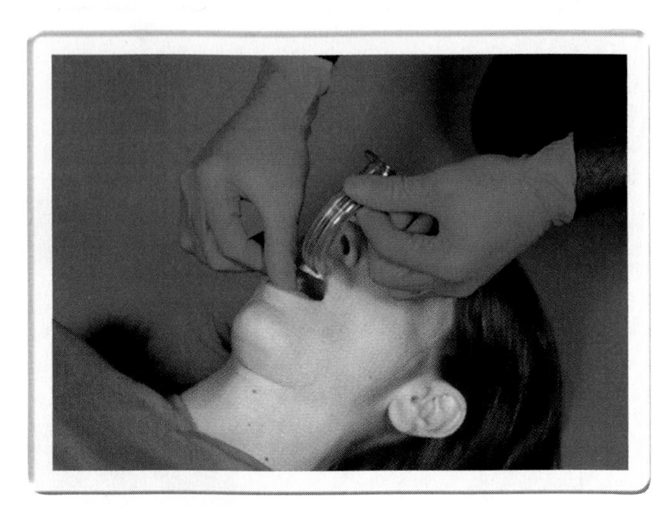

Rotate the airway 180° when no resistance is felt and the airway reaches the soft palate, or insert from the side of the mouth with the tip toward the inside of the cheek.

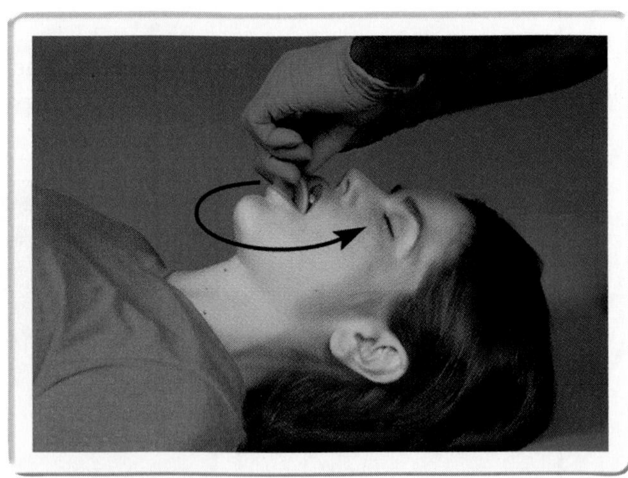

A tongue depressor can be used to hold the tongue inferior and the airway inserted tip down.

Special Populations

Infant Oropharyngeal Airways

In infants, insert the oropharyngeal airway from the side or with the aid of a tongue depressor. Rotating the airway from the top can cause trauma to an infant's palate.

Remove oropharyngeal airway.

If the patient starts to gag, remove the airway by pulling the flange anterior and inferior.

Oral Suctioning

SKILL 8

Performance Objective

Given an adult patient and appropriate equipment, the candidate shall demonstrate proper procedures for performing oral suctioning using the criteria herein prescribed, in 2 minutes or less.

Equipment

The following equipment is required to perform this skill:

- Appropriate body substance isolation/personal protective equipment
- Suction device
- Suction tubing
- Suction catheter
  - Rigid wand (Yankauer type)
  - Flexible catheter

Equipment that may be helpful:

- Adjustable manometer
- Pulse oximeter

Indications

- Removal of blood or liquid emesis from the oropharynx
- Prevention of aspiration of secretions in patients without the ability to protect their own airway

Contraindications

- None

Complications

- Deep suctioning can stimulate the gag reflex, worsening vomiting and promoting bradycardia.
- Suctioning can cause increased intracranial pressure.
- Prolonged suctioning can promote hypoxia and anoxia.
- Incomplete suctioning can lead to forced aspiration in nonbreathing patients ventilated by bag-mask device.

Procedures

Ensure body substance isolation before beginning procedures.

Prior to beginning patient care, appropriate body substance isolation procedures should be employed.

Turn patient's head to side.

Turn the patient's head to the side and begin manual removal of emesis. In the case of cervical spine injury the patient should be turned as a unit, secured to a backboard if possible.

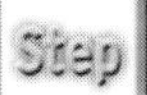 

Turn on and prepare suction device.

Turn on suction device and adjust manometer if available.

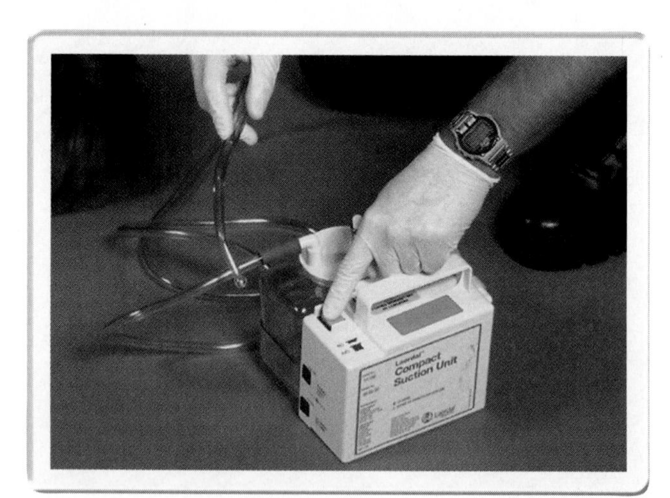

Confirm presence of mechanical suction.

Confirm that suction is working by checking suction and power.

Step 5 Insert suction tip without suction.

Measure the depth of catheter insertion from the earlobe to the corner of the mouth.

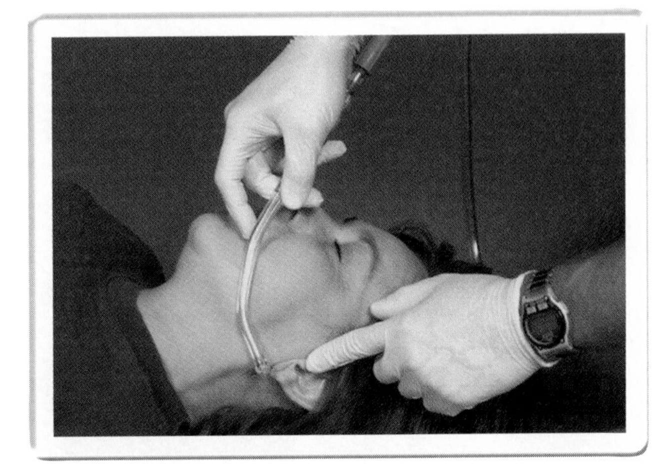

Insert the suction catheter to the proper depth with the suction turned off.

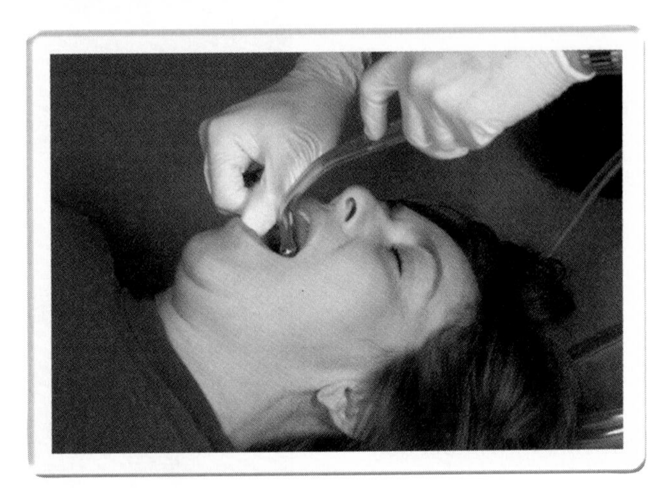

Step 6 Apply suction to the oropharynx.

Apply suction to the catheter. Remove emesis by moving the catheter tip from side to side. Suction the patient for no more than 15 seconds. If the patient is not breathing or is otherwise unable to protect his or her own airway, suction until the airway is clear. Ensure no emesis remains in the nares before providing mechanical ventilation.

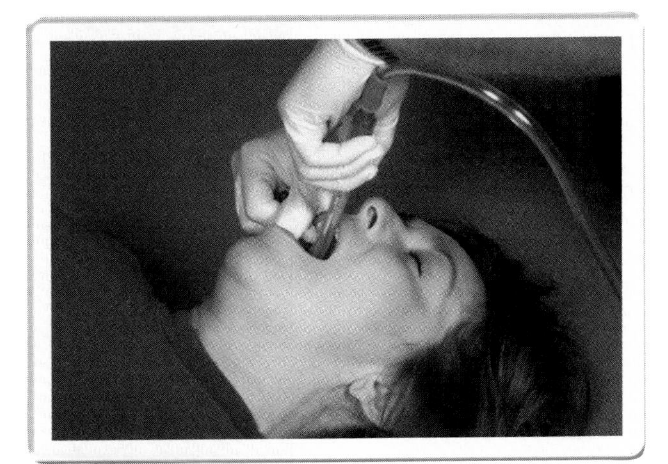

SKILL 9

Nasopharyngeal Airway

Performance Objective

Given an adult patient and appropriate equipment, the candidate shall demonstrate proper procedures for placing a nasopharyngeal airway using the criteria herein prescribed, in 1 minute or less.

Equipment

The following equipment is required to perform this skill:

- Appropriate body substance isolation/personal protective equipment
- Nasopharyngeal airway (various sizes)
- Water-soluble lubricant

Indications

- Basic airway adjunct in patients with intact gag reflexes

Contraindications

- Facial fractures
- Anomalous facial features
- Deviated septum
- Bleeding disorders (use caution)

Complications

- Trauma to the nasal mucosa may cause bleeding and secondary aspiration of blood.
- Nasopharyngeal airways that are too long may cause esophageal intubation or laryngospasm.
- Rarely, vomiting may occur if the gag reflex is stimulated.

Procedures

Step 1 Ensure body substance isolation before beginning procedures.

Prior to beginning patient care, appropriate body substance isolation procedures should be employed.

Step 2 Select appropriate-sized airway.

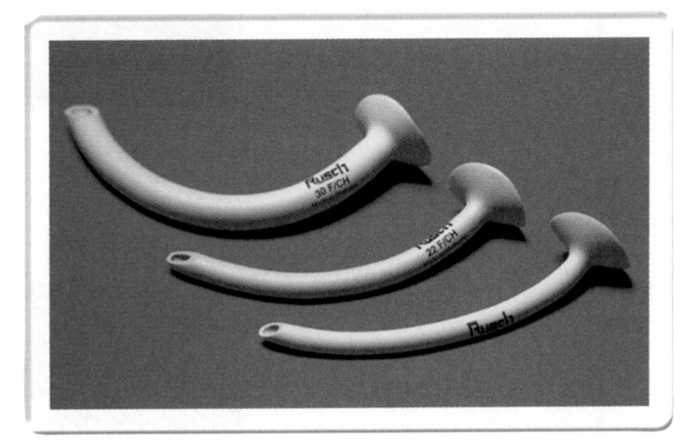

Nasopharyngeal airways are sized to fit the opening of the nose (the nares). Choose an airway that will fit the patient's nares.

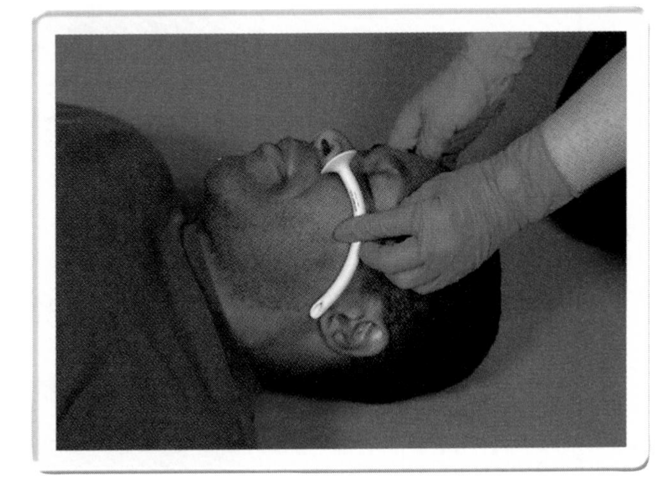

Step 3 Lubricate the nasal airway.

Lubricate the airway with a water-soluble jelly (such as K-Y Jelly) or water-soluble anesthetic jelly (such as Xylocaine jelly).

Step 4 Safely insert the airway with the bevel facing toward the septum.

Insert the nasopharyngeal airway into the chosen nostril with the bevel toward the septum.

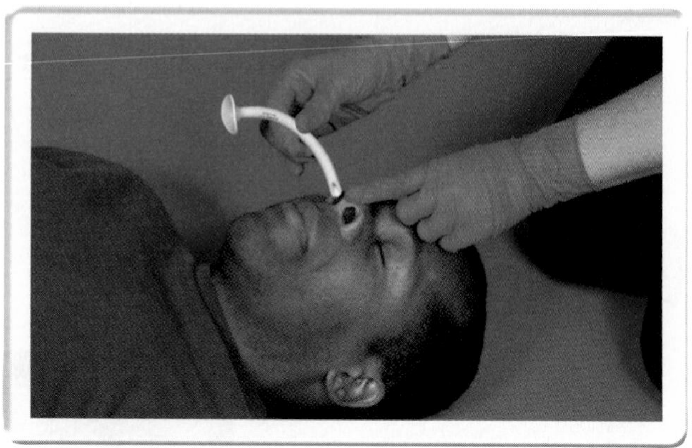

Carefully advance the airway straight down, perpendicular to the face. *Do not* push the airway upward, following the shape of the nose.

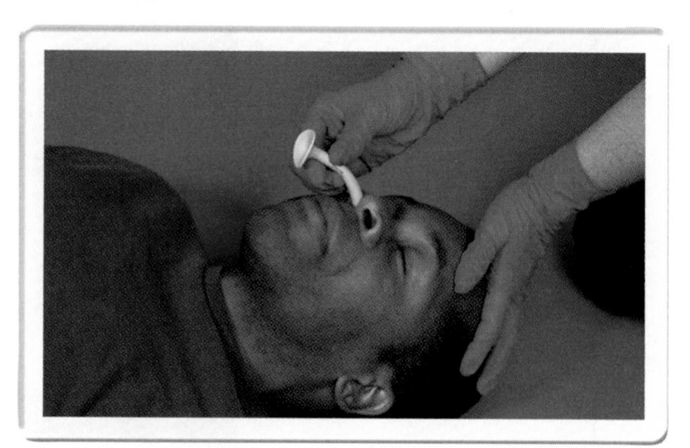

Step 5 Maintain the airway.

Frequently check the airway for proper position and to ensure it has not become occluded with mucus.

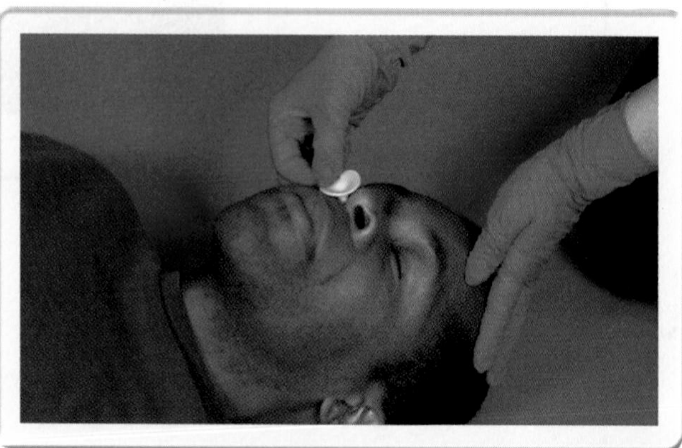

Check Your Knowledge

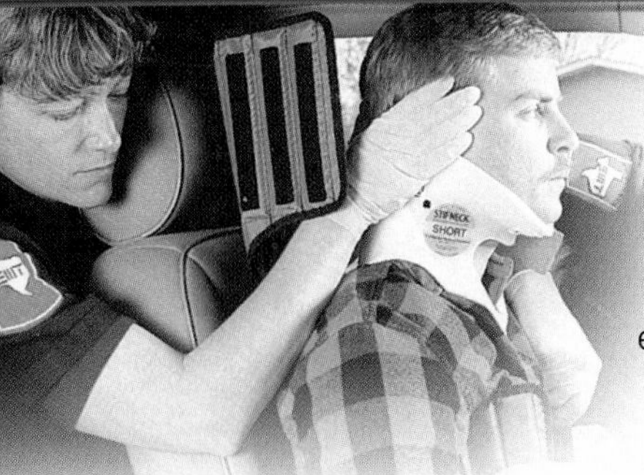

You and your partner are dispatched to a residence on Colonial Drive for an unresponsive female. You arrive to find a family picnic in progress. A 70-year-old woman is lying supine on the patio surrounded by several family members. The family tells you that the patient was sitting at the picnic table when she suddenly slumped over. Two family members lowered her to the ground. You determine that the patient is unresponsive and is not breathing.

1. What method would be most appropriate for opening this patient's airway?
A. Jaw-thrust maneuver
B. Tongue-jaw lift maneuver
C. Head tilt-chin lift maneuver
D. None of the above

2. Put the following steps in the appropriate order for preparing the oxygen tank for delivery of supplemental oxygen:
1. Check tank pressure.
2. Attach oxygen delivery device.
3. Assemble regulator to tank.
4. Open tank.
5. Check for leaks.

A. 1, 2, 3, 4, 5
B. 4, 1, 2, 3, 5
C. 3, 4, 5, 1, 2
D. 4, 5, 3, 1, 2

3. Which type of airway adjunct would you use for this patient?
A. Oropharyngeal airway
B. Nasopharyngeal airway
C. No airway adjunct
D. Oropharyngeal or nasopharyngeal airway

4. To select the proper size oropharyngeal airway, measure from the _________ to the __________.
A. tip of the nose; corner of the mouth
B. tip of the nose; base of the earlobe
C. corner of the mouth; base of the earlobe
D. center of the mouth; base of the earlobe

5. Effective ventilations must be started within how many seconds?
A. 10
B. 20
C. 30
D. 40

6. What oxygen flow rate should be used for supplemental oxygen if provided by a bag-mask device?
A. 6 L/min
B. 10 L/min
C. 12 L/min
D. 15 L/min

7. Ventilations should be delivered at what rate?
A. 8 to 10 breaths/min
B. 10 to 12 breaths/min
C. 12 to 15 breaths/min
D. 10 to 20 breaths/min

8. Each ventilation should last how long?
A. 1 second
B. 2 seconds
C. 5 seconds
D. None of the above

9. If suction is required, you should suction for no longer than _____ seconds.
A. 5
B. 10
C. 15
D. 20

10. Portable oxygen tanks should be replaced when the residual pressure is less than:
A. 200 psi.
B. 500 psi.
C. 1,000 psi.
D. 1,500 psi.

Cardiovascular Emergencies

Introduction

Cardiopulmonary resuscitation (CPR) is the process of using chest compression and ventilation to replace the patient's normal heart and lung functions. The process is intended to ensure minimal blood flow to the brain while other interventions work to return the patient's heartbeat.

Cardiopulmonary resuscitation standards are set by the International Liaison Committee on Resuscitation (ILCOR). The standards reflect the most current thinking and research available for the management of patients in cardiac arrest and in critical cardiac situations.

When assigned the task of performing chest compressions or ventilations, the rescuer must focus all of his or her attention and efforts on the task at hand. The skills in this section will assist you in performing resuscitation techniques for patients of all ages.

SECTION 2

SKILL 10

Adult One-Rescuer CPR

Performance Objective

Given an unconscious adolescent or adult patient, the candidate should begin assessment of the need for CPR and proceed as indicated upon identifying cardiac arrest. The candidate shall perform the procedure for adult CPR in 5 minutes or less.

Equipment

The following equipment is required to perform this skill:

- Appropriate body substance isolation/personal protective equipment

Equipment that may be helpful:

- Oropharyngeal airways (various sizes)
- Nasopharyngeal airways (various sizes)
- Pocket mask with
  - One-way valve
  - Oxygen connecting port
- Bag-mask device
- Oxygen cylinder and regulator
- Automated external defibrillator
- Suction device
- Pulse oximeter
- End-tidal carbon dioxide meter

Indications

- Cardiac arrest in adolescent (ie, displaying secondary sexual characteristics) and adult patients
- Severe bradycardia

Contraindications

- Obvious signs of death
- Valid, verifiable Do Not Resuscitate order

Complications

- Fractured ribs or sternum
- Lacerated liver from fracture of the xiphoid process

Procedures

Step 1 Ensure body substance isolation before beginning procedures.

Prior to beginning patient care, appropriate body substance isolation procedures should be employed.

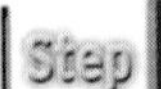

Step 2 Assess level of consciousness.

Grasp the patient's shoulders and gently shake. Firm, nonviolent action may be required to awaken a deeply sleeping or impaired patient. Shout "Are you OK?" or a similar question. If known, use the patient's name. It should take less than 10 seconds to arouse the patient. The patient can be assessed in any position; however, to assess the airway and, more important, to begin chest compressions, the patient should be positioned supine with a hard surface beneath the back.

Step 3 Open airway.

Kneel beside the patient's shoulders and open the airway using the most appropriate method. Begin with the jaw-thrust maneuver until you are sure no cervical spine injury exists. Patients with stiff or large necks may also require a jaw thrust. Once a neck injury has been ruled out, the use of the head tilt–chin lift maneuver is recommended.

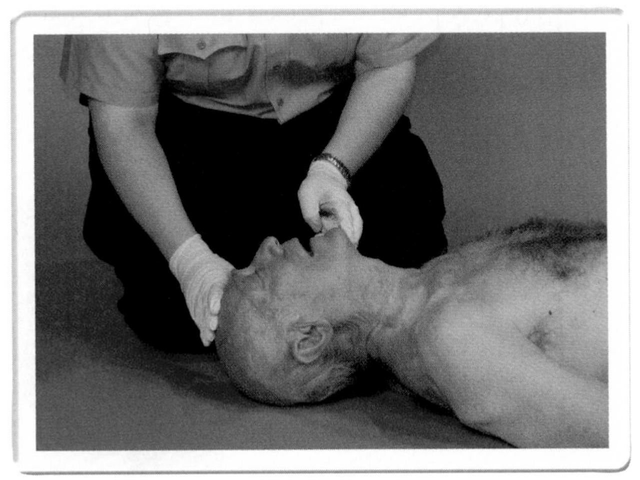

In the Field

The Lone Rescuer: Call for Help or Begin CPR?

As a professional rescuer working as a part of a fully equipped arriving EMS or First Response unit, there is no need to call for help. However, when encountering a patient while you're off duty or when you arrive alone, it is essential to call for help as soon as appropriate. The decision is based on the speed of the patient's collapse. Sudden collapse is most often the result of ventricular fibrillation. In these cases it is important to call for help immediately, because the patient's survival is dependent on early defibrillation. Leave the patient, call for help/defibrillation, and return for further assessment and care. In cases where the cause of collapse is hypoxia or asphyxia, ventilations are essential. Begin CPR and perform five cycles (2 minutes) of chest compressions and ventilations before calling for help.

Step 4 Assessment: Determine breathlessness.

While maintaining the open airway, place your ear approximately 1 inch above the patient's mouth. Face the patient's chest and look for chest rise.

This is known as the *look, listen, and feel* procedure: *Look* for chest rise, *listen* for air exchange, and *feel* for air against your cheek. If any of these are found, the patient is breathing and should be assessed for adequacy of ventilations and other concerns. A breathing patient is usually placed in the recovery position if no trauma is suspected. It should take less than 10 seconds to determine breathlessness.

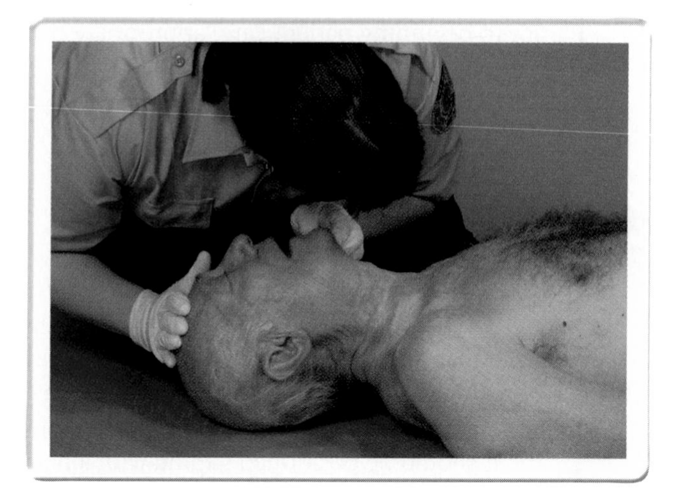

If the victim is breathing, place him or her in the recovery position and continue further assessment.

Step 5 Begin effective ventilations.

Properly position a pocket mask or other barrier device over the patient's face. Deliver two slow breaths, each over 1 full second, causing chest rise. Allow complete deflation of the chest between breaths. It is important to watch the patient's chest throughout the ventilation procedure.

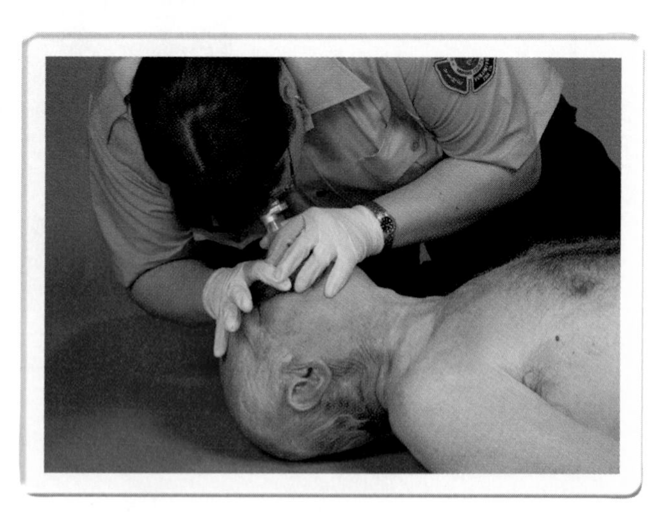

In the Field

Mouth-to-Mouth Ventilations

Although professional rescuers should possess the skills necessary to perform mouth-to-mouth resuscitation, it is not recommended. Instead, professional responders should use mouth-to-mask ventilations or other barrier devices when available.

Step 6 Assessment: Determine pulselessness.

While maintaining an open airway, move your hand under the patient's chin down along the neck and feel for a carotid pulse. It should take less than 10 seconds to assess circulation. If after 10 seconds no definite pulse is felt, prepare to deliver chest compressions. It is very important not to focus on finding a pulse. Delaying the start of compressions can be fatal to the patient.

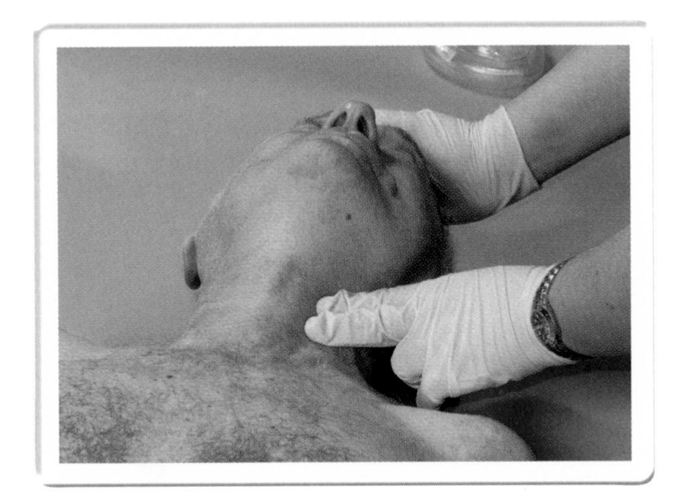

If a pulse is present but the victim is still not breathing, provide rescue breathing, one breath every 5 to 6 seconds. Recheck the victim's pulse about every 2 minutes.

Step 7 Use appropriate hand position for compressions.

Place the heel of one hand over the lower half of the patient's sternum in the center of the chest, between the nipples. Place the heel of the other hand on top of the first.

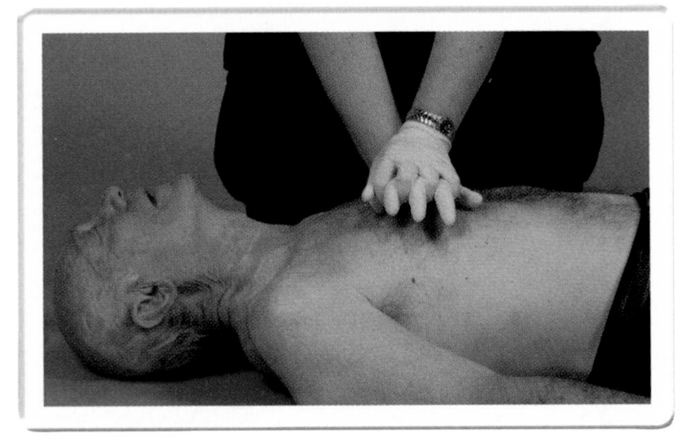

Step 8 Begin chest compressions.

Compress the chest vertically 1.5 to 2 inches, keeping your elbows locked. The compressions should be smooth and even, equally up and down in a fluid motion. The end of the upstroke should allow the chest to return to the relaxed position without removing your hands from the chest. It is often helpful to count out a mnemonic to keep the rhythm.

In the Field

Chest Compressions

Remember to "push hard and push fast." It is very important to the patient that the chest compressions be delivered with absolute effectiveness. Studies of both in-hospital and out-of-hospital compressions have found that 40% of delivered chest compressions are of insufficient depth to develop adequate blood flow. Close attention to the delivery of chest compressions will correct this devastating oversight.

Step 9 Performs steps 1 to 8 in proper sequence.

It is essential that these steps be performed in the proper sequence. Failure to do so may result in chest compressions being performed on an adequately perfusing patient, or ventilations on an adequately breathing patient. Both are embarrassing.

Step 10 Perform five cycles of 30 compressions and two ventilations at a rate of 100/min (5 compressions in 3 to 4 seconds).

Chest compressions are delivered at a rate of 100/min. After each set of 30 compressions, stop, open the airway, and deliver two breaths exactly as described in Step 5. It is essential to relocate the proper hand position on the patient's chest each time chest compressions begin. Continue this sequence of 30 compressions to two ventilations for five cycles, and then stop and reassess the patient for perfusion.

Step 11 Assessment: Determine pulselessness.

Check the pulse as described in Step 6, lasting less than 10 seconds.

Step 12 Continue CPR.

If the patient is still pulseless, resume chest compressions. Relocate the proper hand placement and continue the compression/ventilation ratio of 30:2. Continue to check for signs of perfusion every 5 minutes.

If the patient's pulse returns, the probability that respirations will also return is initially low. Ventilations may need to continue for some time. If respirations do return to an adequate rate and no signs of trauma are found, place the patient in the recovery position and continue to monitor pulse and respirations.

CPR should be continued as long as any chance of survival exists. In advanced life support (ALS) systems, with electrocardiographic monitoring, defibrillation, and ALS capabilities, all resuscitative efforts should be performed at the scene. It should be understood that competent ALS care administered by paramedics in the field provides a greater chance of successful resuscitation than care of the patient transported to the hospital. Studies have shown that there is no chance that a hospital emergency department will be successful in reviving a patient that paramedics could not.

It is also important to accept death. In ALS systems, patients who have been adequately managed but remain in cardiac arrest should not be transported. These patients need resuscitative efforts stopped. However, before deciding to halt efforts, you must be prepared to inform the patient's family that the patient has died. This may be the most difficult step, but it is essential that paramedics learn how to deliver devastating news to a patient's family.

Adult Team CPR

Performance Objective

Given an unconscious adolescent or adult patient and a partner, candidates should begin assessment of the need for CPR and proceed as indicated upon identifying cardiac arrest. The candidates shall perform the procedure for adult team CPR in 5 minutes or less.

Equipment

The following equipment is required to perform this skill:

- Appropriate body substance isolation/personal protective equipment

Equipment that may be helpful:

- Oropharyngeal airways (various sizes)
- Nasopharyngeal airways (various sizes)
- Pocket mask with
  - One-way valve
  - Oxygen connecting port
- Bag-mask device
- Oxygen cylinder and regulator
- Automated external defibrillator
- Suction device
- Pulse oximeter
- End-tidal carbon dioxide meter

Indications

- Cardiac arrest in adolescent (ie, displaying secondary sexual characteristics) and adult patients
- Severe bradycardia

Contraindications

- Obvious signs of death
- Valid, verifiable Do Not Resuscitate order

Complications

- Fractured ribs or sternum
- Lacerated liver from fracture of the xiphoid process

Procedures

Rescuer One takes initial responsibility for patient assessment. Rescuer Two assists Rescuer One.

Step 1 Ensure body substance isolation before beginning procedures.

Prior to beginning patient care, appropriate body substance isolation procedures should be employed.

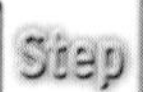

Step 2 Rescuer One: Assess patient.

Following the procedures described in adult one-rescuer CPR (Skill 10), assess the patient's level of consciousness, airway, and ventilations.

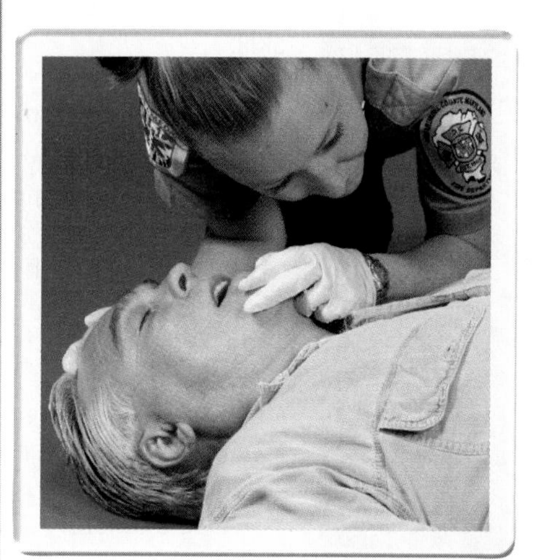

Rescuer Two: Prepare equipment.

Gather and prepare the equipment needed to care for a patient in cardiac arrest. This should include a pocket mask or bag-mask device and an automated external defibrillator (AED). Other equipment that may be helpful includes suction equipment, backboard, and automatic compression devices.

Rescuer One: Instruct partner to begin ventilations and check for pulse.

Upon determining breathlessness, instruct your partner(s) to begin ventilations. Once ventilations have been delivered, assess the patient for the presence of a pulse. Check for a carotid pulse for no more than 10 seconds.

Rescuer Two: Ventilate the patient.

Using a pocket mask or bag-mask device, deliver two ventilations sufficient to cause chest rise. Ventilations should be given over 1 second.

To ensure proper ventilation using a bag-mask device, two rescuers should work together: one to achieve mask seal, the other to squeeze the bag. If a third person is available, cricoid pressure should be applied.

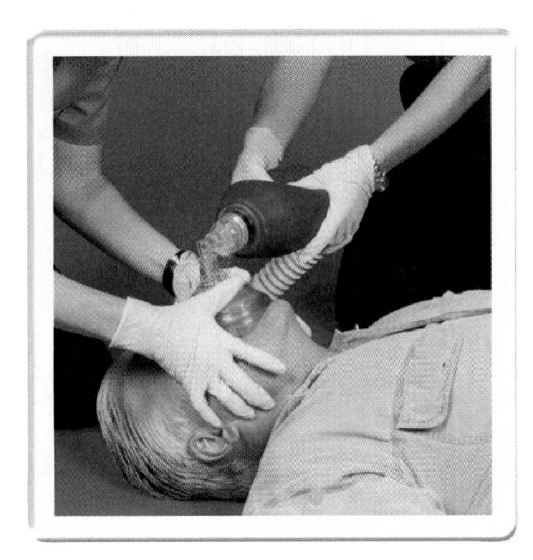

In the Field

Alternate Technique

If only two rescuers are present, it may be more effective for the person at the head to create and hold a two-handed mask seal. The person giving chest compressions can then reach with one hand and deliver two breaths. This will improve ventilation quickly.

Step 4 Rescuer One: Deliver chest compressions.

Locate the correct hand position on the patient's chest and deliver 30 compressions at a rate of 100/min. Repeat compressions after each set of two ventilations for a total of five cycles. In two-rescuer CPR, your hands should never leave the patient's chest, so relocating the correct position between ventilations is not necessary.

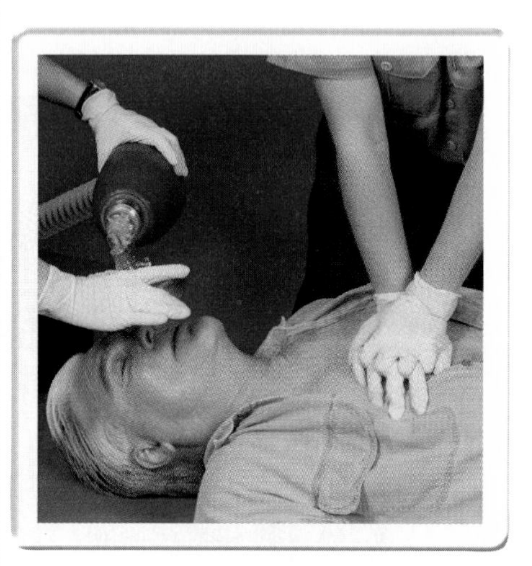

Rescuer Two: Ventilate after each set of 30 compressions. Attach defibrillator pads if available.

Continue to ventilate after each set of 30 compressions for a total of five cycles. Between sets of ventilations, attach defibrillator pads to the patient's chest if they are available.

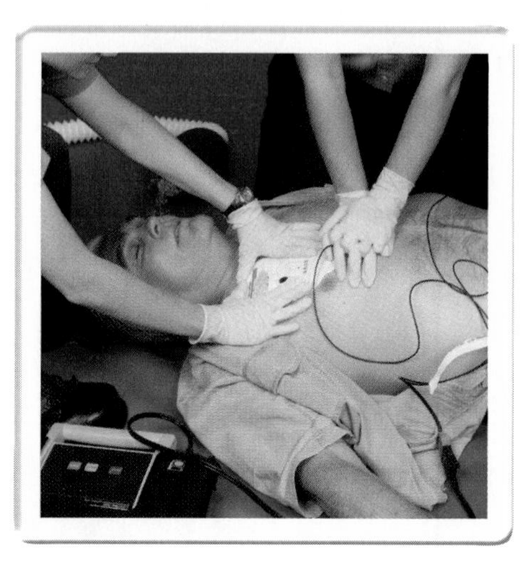

Step 5 Rescuer One: Reassess patient.

Immediately after giving the last set of chest compressions, reassess the patient for the presence of a pulse. If an AED or standard defibrillator is available, check for the presence of a shockable rhythm. If a shockable rhythm is identified, deliver the required energy level.

Rescuer Two: Move to chest.

Move to the patient's chest and prepare for chest compressions. If an AED or standard defibrillator has been attached, stand clear of the patient and avoid patient movement while the rhythm is being identified.

Rescuer One: Instruct partner to begin compressions.

Immediately after determining the absence of a pulse, or immediately after the defibrillator has delivered its shock, instruct your partner to begin chest compressions.

Rescuer Two: Begin chest compressions.

Locate the correct hand position on the patient's chest and deliver 30 compressions at a rate of 100/min.

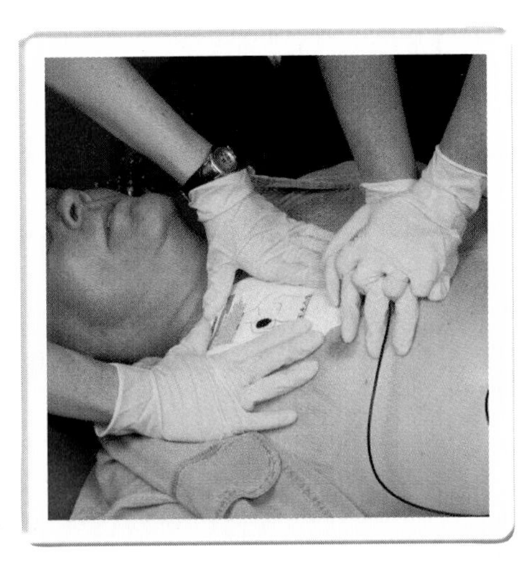

Rescuer One: Ventilate the patient after each set of 30 compressions.

Continue to ventilate after each set of 30 compressions for a total of five cycles.

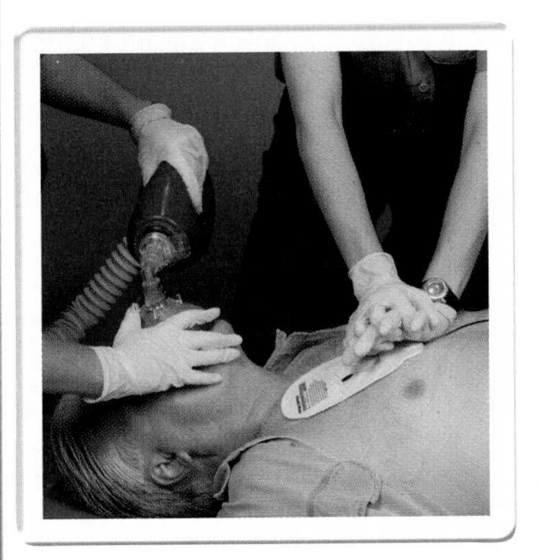

Rescuer Two: Repeat compressions for a total of five cycles.

Repeat compressions after each set of two ventilations for a total of five cycles. In two-rescuer CPR, your hands should never leave the patient's chest, so relocating the correct position between ventilations is not necessary.

In the Field

Adequate Ventilations

Patients experiencing sudden cardiac arrest may display occasional gasps. These should not be considered adequate ventilations. Treat the patient with occasional gasps as if he or she were not breathing.

Special Populations

CPR Following Endotracheal Intubation

A patient who is intubated with a properly placed endotracheal tube can have chest compressions performed simultaneously with ventilations, removing the need to pause. The compression-to-ventilation ratio no longer exists. Chest compressions should be given uninterrupted at a rate of 100/min, while ventilations are given at a rate of 6 to 8 per minute. Remember to prevent compressor fatigue and a reduction in compression rate and depth by rotating the responsibilities every 2 minutes.

In the Field

Interposed Abdominal Compressions

If a third rescuer is available, interposed abdominal compressions can be delivered. This technique is an adjunct to chest compressions and is used to enhance venous return during CPR. The technique is performed by compressing the abdomen midway between the xiphoid process and the umbilicus during the relaxation phase of the chest compressions. Properly performed, the chest and abdominal compressions should function like a see-saw, with one up and one down at any one time.

Although there is little evidence to recommend abdominal compressions in the prehospital setting, there is no reason the technique should not be employed if the personnel are properly trained and it is approved by medical control.

Step 8 Reassess patient.

After each set of five cycles, assess the patient for the presence of a pulse. If an AED or standard defibrillator is available, assess the patient for a shockable rhythm. If a shockable rhythm is identified, deliver the required energy level.

An effective switch should have both rescuers repositioned and CPR reestablished in less than 10 seconds. It is always best, when possible, for rescuers to work on opposite sides of the patient. This allows rescuers to simply move up or down during the switch. However, in the back of an ambulance it may be necessary for the compressor to move forward into the door opening while the ventilator slides down to the chest. This allows for a smooth switch without crawling on top of each other.

Step 9 Continue compressions and ventilations.

Continue to deliver compressions and ventilations as a team. Remember to change responsibilities for chest compressions with each assessment of the patient.

Step 10 Prepare patient for transport and advanced life support.

While maintaining the sequence of chest compressions and ventilations, prepare the patient for transport. The decision of when and how to accomplish this move should not be taken lightly. Basic life support units should work toward transport from the first minute they arrive on scene. Advanced life support units should only transport after all initial patient care procedures and medications have been accomplished or when the patient will benefit more from rapid transport than from on-scene patient care. It is important to remember that CPR in the back of a moving ambulance, especially an ambulance that is being driven in an emergency mode, is less than optimal. CPR in the back of a moving ambulance is also hazardous to the personnel providing patient care.

When advanced life support is available, endotracheal intubation will be established in most cases. Once the patient is intubated, chest compressions should be given at a rate of 100/min and without stopping for the ventilations. Ventilations will then be delivered at a rate of 6 to 8 ventilations per minute using a bag-mask device.

Adult Foreign Body Airway Obstruction

SKILL 12

Performance Objective

Given a conscious adolescent or adult patient who appears to be choking, the candidate should begin assessment of the need to perform abdominal thrusts and proceed as indicated upon identifying an airway obstruction. The candidate shall perform the procedure for conscious adult foreign body airway obstruction in 3 minutes or less.

Equipment

The following equipment is required to perform this skill:

- Appropriate body substance isolation/personal protective equipment
- Pocket mask with
  - One-way valve
  - Oxygen connecting port

Equipment that may be helpful:

- Oropharyngeal airways (various sizes)
- Nasopharyngeal airways (various sizes)
- Bag-mask device
- Oxygen cylinder and regulator
- Automated external defibrillator
- Suction device
- Pulse oximeter
- End-tidal carbon dioxide meter

Indications

- Foreign body airway obstruction in the adolescent (ie, displaying secondary sexual characteristics) or adult patient

Contraindications

- Patients who can speak or cough

Complications

- None if properly performed

Procedures

Step 1 Ensure body substance isolation before beginning procedures.

Prior to beginning patient care, appropriate body substance isolation procedures should be employed.

Step 2 Assess airway and breathing.

Begin your assessment with the simple question "Are you choking?" If the patient nods "yes" or continues to grab his or her throat and does not speak, prepare to perform abdominal thrusts to remove the obstruction from the airway.

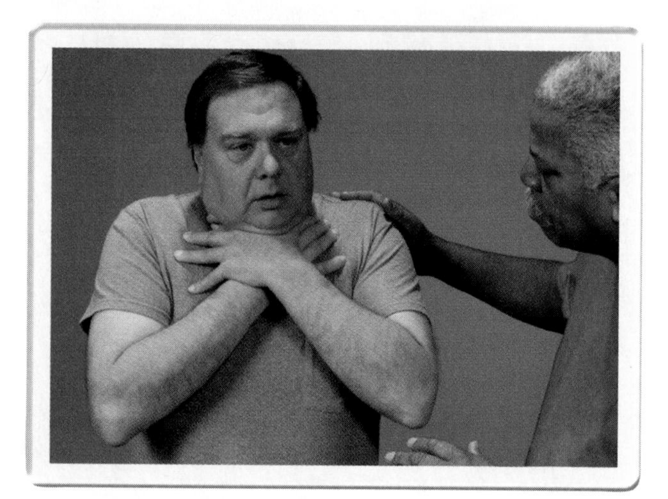

If the patient is able to breathe on his or her own and can speak, the patient is able to move air and does not need emergent assistance. Encourage the patient to cough. Inform the patient you are going to help him or her. Continue to talk to the patient and explain each step as you perform the procedure.

Step 3 Stand behind the patient and position hands.

Position yourself behind the patient. Reach around the patient's waist and make a fist with one hand.

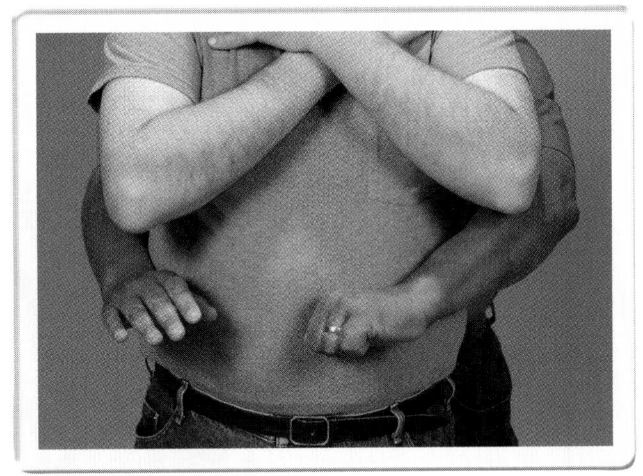

With the thumb side against the patient's abdomen, place the fist just above the umbilicus, but below the xiphoid process. Grasp your fist with your opposite hand.

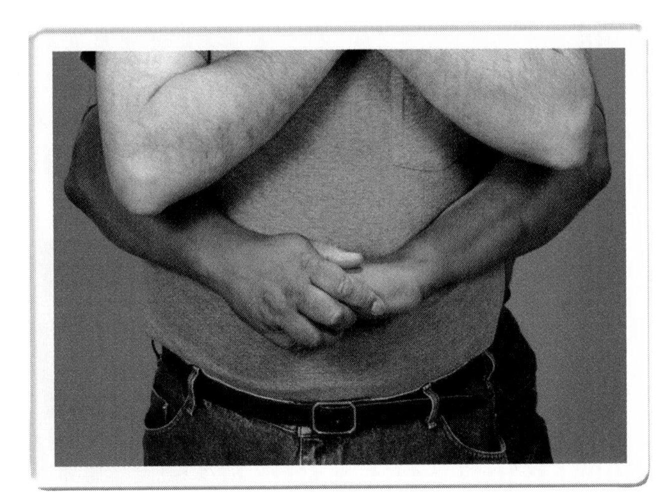

Step 4 Perform abdominal thrusts.

With both hands in position, apply quick, upward thrusts into the patient's abdomen. Ensure each thrust is individual and distinct, allowing the abdomen to return to its original position.

Step 5 Repeat abdominal thrusts.

Repeat thrusts until the foreign body is expelled, the patient can breathe spontaneously, or the patient becomes unconscious.

If the patient becomes unconscious, guide the patient to the ground and call for help.

Special Populations

Foreign Body Airway Maneuvers on Large Patients

Patients who are obese or pregnant, or who in general have a large abdominal mass, cannot have abdominal thrusts performed with success. In these patients you can achieve the same effect through the use of chest compressions. Chest compressions are delivered by standing behind the patient and placing your arms around his or her chest and delivering thrusts to the midsternum between the nipple line. If the patient is too large for you to reach around his or her chest, consider moving the patient against a wall and performing the compressions from the front. Remember to duck.

Step 6 Deliver chest compressions (for unconscious patients or for patients who become unconscious while removing an airway obstruction).

Place the heel of one hand over the lower half of the patient's sternum in the center of the chest, between the nipples.

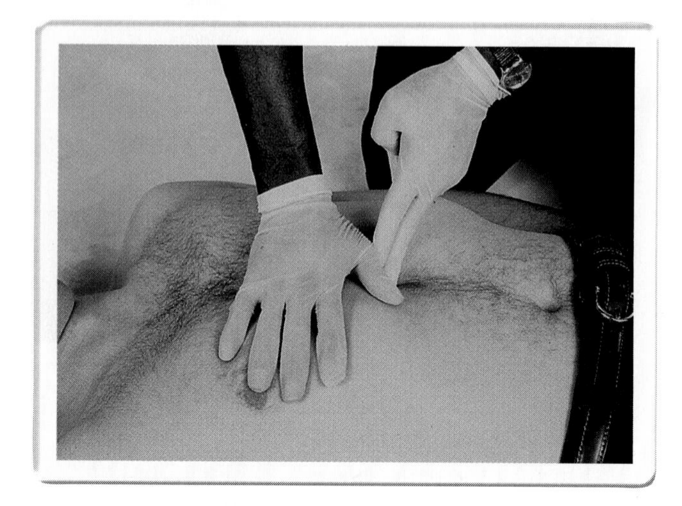

Place the heel of the other hand on top of the first.

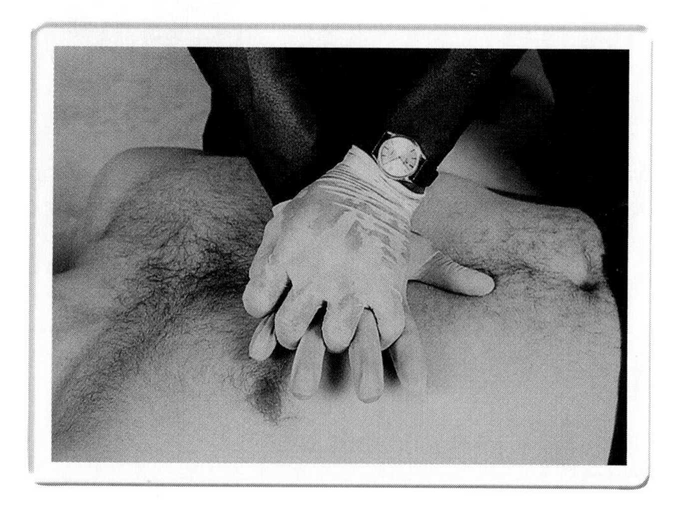

Compress the chest vertically 1.5 to 2 inches, keeping your elbows locked. The compressions should be smooth and even, equally up and down in a fluid motion. The end of the upstroke should allow the chest to return to the relaxed position without removing your hands from the chest. Deliver the full 30 compressions.

Step 7 Inspect the airway and assess for breathing.

After delivering 30 compressions, open the patient's mouth and look for the dislodged foreign object. If it is seen, perform a finger sweep to remove the object.

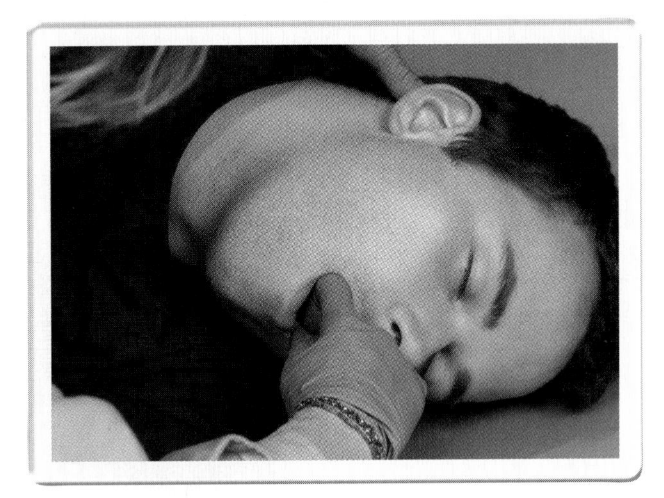

If the object cannot be removed, attempt to ventilate around it rather than waste time in recovery. If no object is seen, attempt to ventilate. If able to ventilate, give two breaths, each lasting 1 full second.

Step 8 Repeat steps 6 and 7 until successful.

Continue with chest compressions, inspection, and attempted ventilations until ventilations are effective. If the object is not able to be recovered, it may be necessary for a person with appropriate training to remove it with direct visualization and Magill forceps. This is an advanced procedure, but it is the definitive treatment.

SKILL 13

Child CPR

Performance Objective

Given an unconscious child patient, the candidate should begin assessment of the need for CPR and proceed as indicated upon identifying cardiac arrest. The candidate shall perform the procedure for child CPR in 5 minutes or less.

Equipment

The following equipment is required to perform this skill:

- Appropriate body substance isolation/personal protective equipment

Equipment that may be helpful:

- Oropharyngeal airways (various sizes)
- Nasopharyngeal airways (various sizes)
- Bag-mask device
- Oxygen cylinder and regulator
- Automated external defibrillator
- Suction device
- Pulse oximeter
- End-tidal carbon dioxide meter

Indications

- Cardiac arrest in children between the ages of 1 to adolescence (ie, displaying secondary sexual characteristics)
- Severe bradycardia

Contraindications

- Obvious signs of death
- Valid, verifiable Do Not Resuscitate order

Complications

- Fractured ribs or sternum
- Lacerated liver from fracture of the xiphoid process

Procedures

 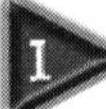

Ensure body substance isolation before beginning procedures.

Prior to beginning patient care, appropriate body substance isolation procedures should be employed.

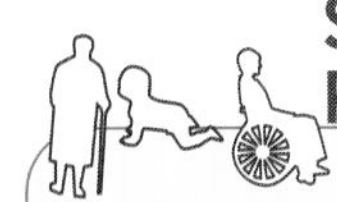

Special Populations

The Child Patient

For the purposes of basic cardiac life support, ILCOR recommends that professional health care providers consider a child any patient who is older than 1 year and has not developed secondary sexual characteristics. This extends the definition given to the lay rescuer from 8 to nearly 14 years of age.

Determine unresponsiveness and position patient.

Grasp the patient's shoulders and gently shake. Firm, nonviolent action may be required to awaken a deeply sleeping or impaired patient. Shout "Are you OK?" or a similar question. If possible, use the patient's name. It should take no more than 10 seconds to arouse the patient. The patient can be assessed in any position; however, to assess the airway and, more important, to begin chest compressions, the patient should be positioned supine with a hard surface beneath the back.

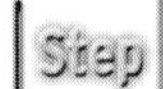 

Open airway.

Kneel beside the patient's shoulders and open the airway using the most appropriate method. Patients with potential cervical injury will require the use of a jaw-thrust maneuver, but in most circumstances the head tilt–chin lift maneuver is recommended. Position your hand nearest the patient's head on the patient's forehead. Place the fingers of your other hand on the underside of the patient's chin. Pulling up on the chin and pushing down on the forehead will tilt the head back and open the airway. The head should rest between 45° and 60° back. Children will require less manipulation than adults.

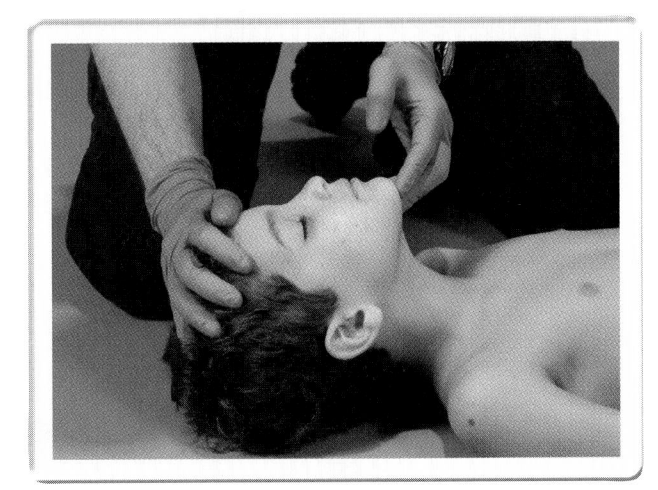

Special Populations

The Lone Rescuer: Call for Help or Begin CPR?

Unlike adults, children seldom experience ventricular fibrillation. Most cases of cardiac arrest are the result of a serious respiratory compromise. In most cases, children experience cardiac arrest following a hypoxic episode. It is important to administer CPR with five cycles (2 minutes) of compressions and ventilations before calling for help. When the cardiac arrest does appear suddenly, it is important to call for help first, and then return to the patient and begin CPR.

 **Step 4** Assessment: Determine breathlessness.

While maintaining the open airway, place your ear approximately 1 inch above the patient's mouth. Face the patient's chest and look for chest rise. This is known as the *look, listen, and feel* procedure: *Look* for chest rise, *listen* for air exchange, and *feel* for air against your cheek. If any of these are found and no trauma is suspected, the patient is breathing and should be placed in the recovery position and assessed for adequacy of ventilations and other concerns. It should take no more than 10 seconds to determine breathlessness. Remember that in children, small chest movement and more subtle breaths will be present. You must pay close attention to all signs.

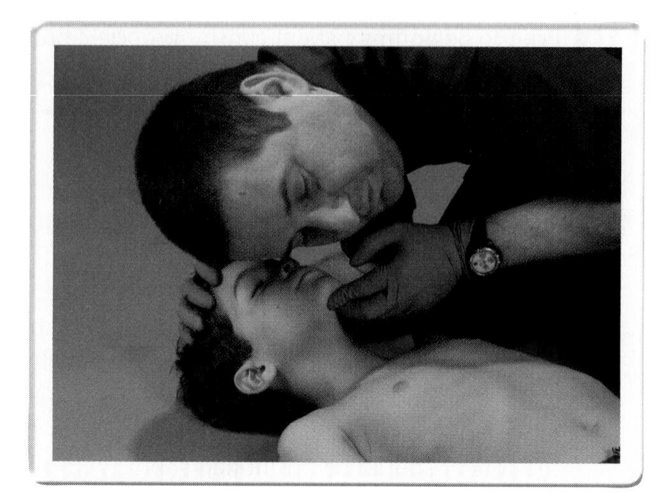

  **Step 5** Ventilate twice.

While maintaining the open airway, pinch the patient's nose and make a tight seal over the patient's mouth with a barrier device.

In smaller children it may be possible, and necessary, to occlude the nose with your cheek. Deliver two gentle breaths, each lasting 1 second, to cause chest rise. Allow complete deflation of the chest between breaths. It is important to watch the patient's chest throughout the ventilation procedure.

Step 6 Assessment: Determine pulselessness.

Slide the hand under the chin down along the patient's neck and feel for a carotid pulse. It should take no more than 10 seconds to assess circulation. If there are no signs of definitive circulation present after 10 seconds, prepare to deliver chest compressions. It is very important not to focus on finding a pulse. In a patient who is not breathing, is unconscious, and not moving, consider the pulse to be absent. Delaying the start of compressions can be fatal to the patient.

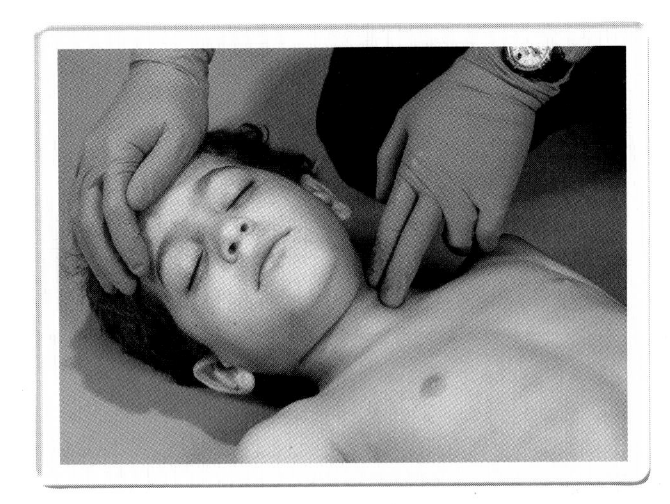

If a pulse is present but the victim is still not breathing, provide rescue breathing, one breath every 3 to 5 seconds. Recheck the victim's pulse about every 2 minutes.

Step 7 Use appropriate hand position for compressions.

Place the heel of your hand on the lower half of the patient's sternum between the nipples. In larger children it may be necessary to use your second hand as you would on the adult patient.

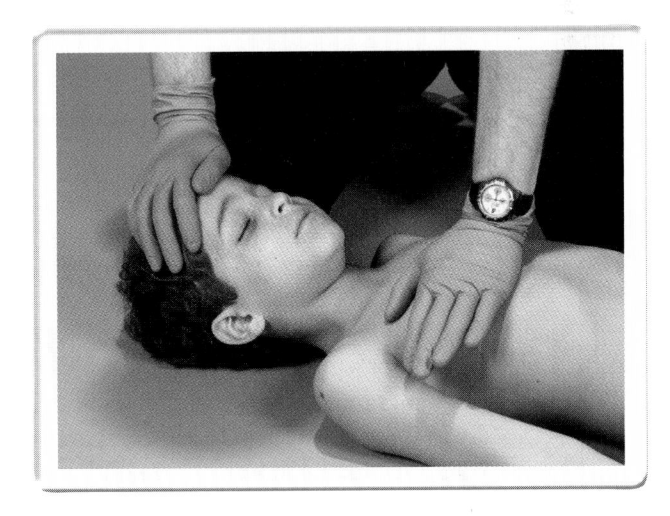

Step 8 Begin chest compressions.

Compress the chest vertically to one third to one half the depth of the patient's chest, keeping your elbow locked. The compressions should be smooth and even, equally up and down in a fluid motion. The end of the upstroke should allow the chest to return to the relaxed position without removing your hands from the chest. It is often helpful to count out a mnemonic to keep the rhythm.

Step 9 Perform steps 1 to 8 in proper sequence.

It is essential that these steps be performed in the proper sequence. Failure to do so may result in chest compressions being performed on an adequately perfusing patient, or ventilations on an adequately breathing patient.

Step 10 Perform five cycles of 30 compressions and two ventilations at a rate of 100/min (5 compressions in 3 to 4 seconds).

Chest compressions are delivered at a rate of 100/min. After each set of 30 compressions, stop, open the airway, and deliver two breaths exactly as performed in Step 5. Continue this sequence of 30 compressions to two ventilations for five cycles, and then stop and reassess the patient for perfusion.

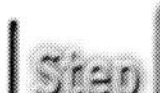
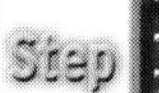

Step 11 Reassessment: Determine pulselessness.

Check the pulse as described in Step 6, lasting no more than 10 seconds.

Special Populations

Working as a Team

When professional rescuers are working as a team in the care of a child or infant patient in cardiac arrest, it is recommended that 15 compressions be given for every two ventilations. Once the child has been intubated, no pause between compressions is necessary, but be sure to give compressions at a rate of 100/min and to ventilate at 12 to 20 breaths/min.

Step 12 Continue CPR.

If the patient is still pulseless, resume chest compressions. Relocate the proper hand placement and continue the compression/ventilation ratio of 30:2. Continue to check for signs of perfusion every five cycles.

If the patient's pulse returns, the probability that respirations will also return is initially low, but much better than compared with an adult. Ventilations may need to continue for some time. If respirations do return to an adequate rate and no signs of trauma are found, place the patient in the recovery position and continue to monitor pulse and respirations.

CPR should continue as long as any chance of survival exists. In advanced life support (ALS) systems, with electrocardiographic monitoring, defibrillation, and ALS capabilities, all resuscitative efforts should be performed at the scene. It should be understood that competent ALS care administered by paramedics

continued

in the field provides a greater chance of successful resuscitation than care of the patient transported to the hospital.

It is also important to accept death. In ALS systems, patients who have been adequately managed but remain in cardiac arrest should not be transported. These patients need resuscitative efforts stopped. However, before deciding to halt efforts, you must be prepared to inform the patient's family that the patient has died. This may be the most difficult step, but it is essential that paramedics learn how to deliver devastating news to a pediatric patient's family. If transport is determined to be necessary, it is important that a pediatric facility be considered rather than a local or regional hospital. Pediatric care is very specialized. Transporting to local hospitals is not in the patient's best interest and in many ways could be considered negligence.

SKILL 14

Child Foreign Body Airway Obstruction

Performance Objective

Given a conscious child patient who appears to be choking, the candidate should begin assessment of the need to perform abdominal thrusts and proceed as indicated upon identifying an airway obstruction. The candidate shall perform the procedure for conscious child foreign body airway obstruction in 3 minutes or less.

Equipment

The following equipment is required to perform this skill:

- Appropriate body substance isolation/personal protective equipment

Equipment that may be helpful:

- Oropharyngeal airways (various sizes)
- Nasopharyngeal airways (various sizes)
- Bag-mask device
- Oxygen cylinder and regulator
- Automated external defibrillator
- Suction device
- Pulse oximeter
- End-tidal carbon dioxide meter

Indications

- Foreign body airway obstruction in children between the ages of 1 to adolescence (ie, displaying secondary sexual characteristics)

Contraindications

- Patients who can speak or cough

Complications

- None if properly performed

Procedures

Ensure body substance isolation before beginning procedures.

Prior to beginning patient care, appropriate body substance isolation procedures should be employed.

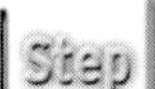 

Assess airway and breathing.

Begin your assessment with the simple question "Are you choking?" If the child nods "yes" or continues to grab her or his throat and does not speak, prepare to perform abdominal thrusts to remove the obstruction from the airway.

If the child is able to breathe on his or her own and can speak, the patient is able to move air and does not need emergent assistance. Encourage the child to cough. Inform the child you are going to help him or her. Continue to talk to the child and explain each step as you perform the procedure.

Stand behind the child and position hands.

Position yourself behind the child. Reach around the child's waist and make a fist with one hand. With the thumb side against the child's abdomen, place the fist just above the umbilicus, but below the xiphoid process. Grasp your fist with your opposite hand.

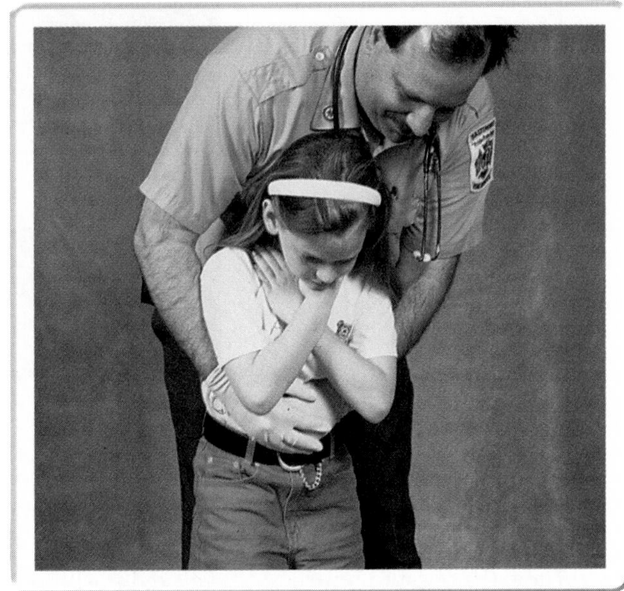

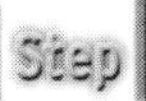 

Step 4 Perform abdominal thrusts.

With both hands in position, apply quick, upward thrusts into the child's abdomen. Ensure each thrust is individual and distinct, allowing the abdomen to return to its original position.

Step 5 Repeat abdominal thrusts.

Repeat thrusts until the foreign body is expelled, the child can breathe spontaneously, or the child becomes unconscious.

If the child becomes unconscious, guide the child to the ground and call for help.

Step 6 Start CPR (for unconscious patients or for patients who become unconscious while removing an airway obstruction).

Attempt to ventilate the patient. If unsuccessful, reopen the airway and again attempt ventilation.

Perform chest compressions. Place the heel of your hand on the lower half of the child's sternum between the nipples. In larger children it may be necessary to use your second hand as you would on the adult patient.

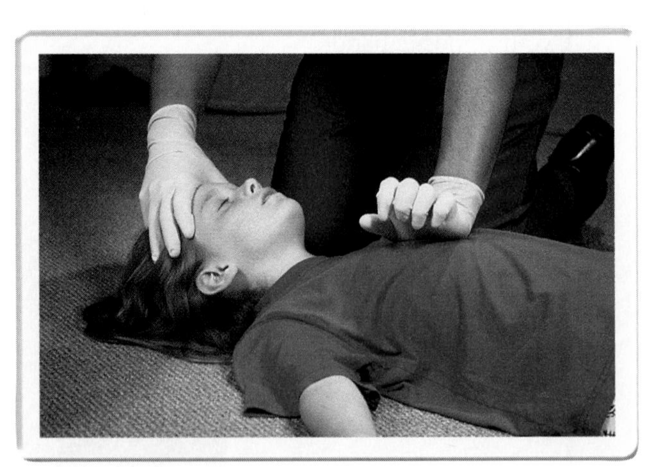

Compress the chest vertically to one third to one half the depth of the child's chest, keeping your elbow locked. The compressions should be smooth and even, equally up and down in a fluid motion. The end of the upstroke should allow the chest to return to the relaxed position without removing your hands from the chest. Deliver the full 30 compressions.

Step 7 Inspect the airway and assess for breathing.

After delivering 30 compressions, open the child's mouth and look for the dislodged foreign object. If it is seen, perform a finger sweep to remove the object. If the object cannot be removed, attempt to ventilate around it rather than waste time in recovery. If no object is seen, attempt to ventilate. If able to ventilate, give two breaths, each lasting 1 full second.

Step 8 Repeat steps 6 and 7 until successful.

Continue with compressions, inspection, and attempted ventilations until ventilations are effective. If the object is not able to be recovered, it may be necessary for a person with appropriate training to remove it with direct visualization and Magill forceps. This is an advanced procedure, but it is the definitive treatment.

SKILL 15 Infant CPR

Performance Objective

Given an unconscious infant, the candidate should begin assessment of the need for CPR and proceed as indicated upon identifying cardiac arrest. The candidate shall perform the procedure for infant CPR in 5 minutes or less.

Equipment

The following equipment is required to perform this skill:

- Appropriate body substance isolation/personal protective equipment

Equipment that may be helpful:

- Oropharyngeal airways (various sizes)
- Nasopharyngeal airways (various sizes)
- Bag-mask device
- Oxygen cylinder and regulator
- Suction device
- Pulse oximeter
- End-tidal carbon dioxide meter

Indications

- Cardiac arrest in patients younger than 1 year
- Bradycardia with associated poor perfusion

Contraindications

- Obvious signs of death
- Valid, verifiable Do Not Resuscitate order

Complications

- Fractured ribs or sternum
- Lacerated liver from fracture of the xiphoid process

Procedures

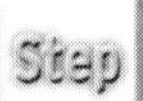 

Step 1 Ensure body substance isolation before beginning procedures.

Prior to beginning patient care, appropriate body substance isolation procedures should be employed.

Step 2 Determine unresponsiveness and position patient.

Gently grasp the victim and call out loudly, "Are you all right?" Tapping or flicking the infant's feet may also be used. Firm, nonviolent action may be required to awaken a deeply sleeping or impaired infant. If known, use the infant's name. It should take no more than 10 seconds to arouse the infant. The infant can be assessed in any position; however, to assess the airway and, more important, to begin chest compressions, the infant should be positioned supine with a hard surface beneath the back.

Step 3 Open airway.

Open the airway using a head tilt–chin lift maneuver. An infant's airway is more pliable than that of adults; thus, less extension is needed. Overextension can actually occlude the airway.

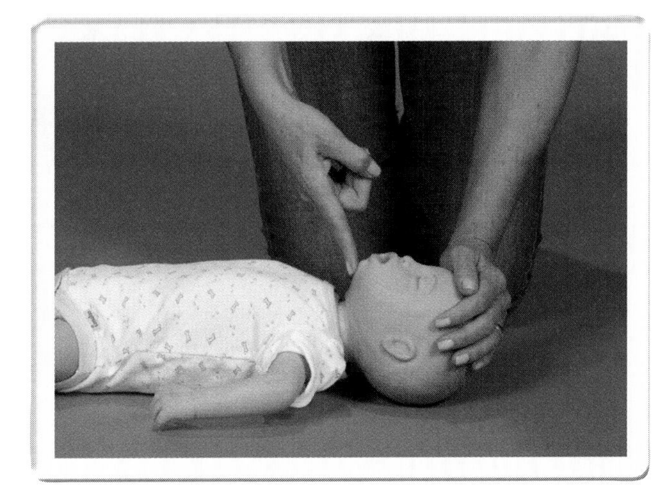

Step 4 Assessment: Determine breathlessness.

While maintaining the open airway, place your ear approximately 1 inch above the infant's mouth. Face the infant's chest and look for chest rise. This is known as the *look, listen, and feel* procedure: *Look* for chest rise, *listen* for air exchange, and *feel* for air against your cheek. If any of these are found, the infant is breathing and should be assessed for adequacy of ventilations and other concerns. It should take no more than 10 seconds to determine breathlessness. Remember that in infants and children, small chest movement and more subtle breaths will be present. You must pay close attention to all signs.

Step 5 Ventilate twice.

While maintaining the open airway, cover both the infant's mouth and nose with your mouth. Deliver two slow breaths, each lasting 1 full second, causing adequate chest rise. Allow complete deflation of the chest between breaths. It is important to watch the infant's chest throughout the ventilation procedure.

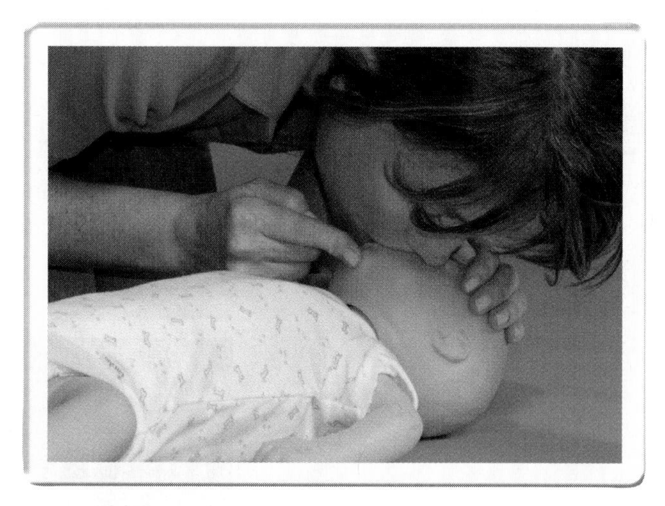

Step 6 Assessment: Determine pulselessness.

While maintaining an open airway, move the hand from under the chin down along the infant's arm and feel for a brachial pulse.

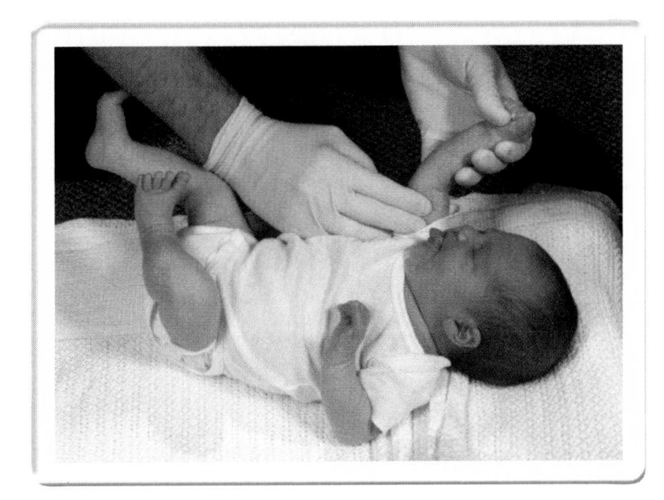

It should take no more than 10 seconds to assess circulation. If after 10 seconds no signs of circulation are present, prepare to deliver chest compressions (see Steps 7 and 8). It is very important not to focus on finding a pulse. In an infant who is not breathing, not conscious, and not moving, consider the pulse to be absent. Delaying the start of compressions can be fatal to the infant.

If a pulse is present but the infant is still not breathing, provide rescue breathing, one breath every 3 to 5 seconds. Recheck the victim's pulse about every 2 minutes.

Step 7 Use appropriate finger position for compressions.

Locate the correct position on the chest by placing two fingers just below the midnipple line.

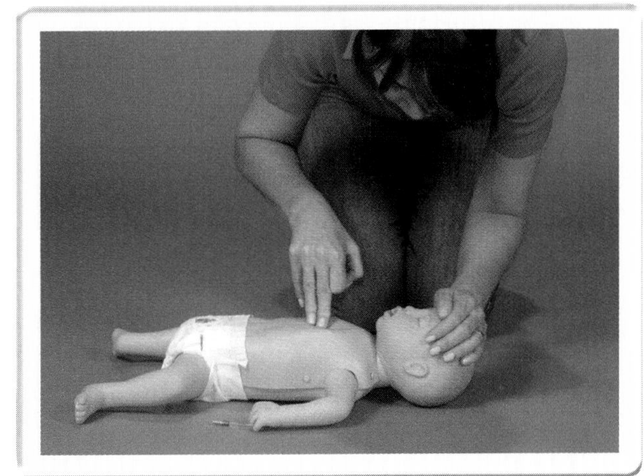

In larger infants, it may be necessary to use three fingers to adequately depress the sternum. When using three fingers, position the first two as just described and place the third finger at the midnipple line.

Step 8 Begin chest compressions.

Using the fingers, compress vertically one third to one half the depth of the chest. Make sure that you achieve equal compression and relaxation cycles. Without removing the fingers from the chest, allow complete chest relaxation on upstroke.

Step 9 Perform steps 1 to 8 in proper sequence.

It is essential that these steps be performed in the proper sequence. Failure to do so may result in chest compressions being performed on an adequately perfusing infant, or ventilations on an adequately breathing infant.

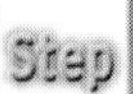 

Step 10 Perform five cycles of 30 compressions and two ventilations at a rate of at least 100/min.

Chest compressions are delivered at a rate of 100/min. After each set of 30 compressions, stop and deliver two breaths exactly as performed in Step 5. Since the fingers should never leave the infant's chest, it is not necessary to relocate the proper finger position each time chest compressions begin. Continue this sequence of 30 compressions to two ventilations for five cycles, and then stop and reassess the infant for perfusion.

Step 11 Reassessment: Determine pulselessness.

Check the pulse as described in Step 6, lasting no more than 10 seconds.

Step 12 Continue CPR.

If the infant is still pulseless, resume chest compressions. Relocate the proper finger placement and continue the compression/ventilation ratio of 30:2. Continue to check for signs of perfusion every 5 minutes.

If the infant's pulse returns, the probability that respirations will also return is low initially, but much better than compared with an adult. Ventilations may need to continue for some time. If respirations do return to an adequate rate and no signs of trauma are found, place the infant in the recovery position and continue to monitor pulse and respirations.

CPR should continue as long as any chance of survival exists. In advanced life support (ALS) systems, with electrocardiographic monitoring, defibrillation, and ALS capabilities, all resuscitative efforts should be performed at the scene.

continued

It should be understood that competent ALS care administered by paramedics in the field provides a greater chance of successful resuscitation than care of the patient transported to the hospital.

It is also important to accept death. In ALS systems, patients who have been adequately managed but remain in cardiac arrest should not be transported. These patients need resuscitative efforts stopped. However, before deciding to halt efforts, you must be prepared to inform the patient's family that the patient has died. This may be the most difficult step, and it may be better to transport and have the news of death relayed by a physician. Almost all pediatric arrests are transported to an appropriate medical facility. It is important that a pediatric trauma center be considered rather than a local or regional hospital because pediatric care is very specialized. Stabilization at local hospitals is not possible in most cases.

SKILL 16

Infant Foreign Body Airway Obstruction

Performance Objective

Given a conscious infant who appears to be choking, the candidate should begin assessment of the need to perform back slaps and chest thrusts, and proceed as indicated upon identifying an airway obstruction. The candidate shall perform the procedure for conscious infant foreign body airway obstruction in 3 minutes or less.

Equipment

The following equipment is required to perform this skill:

- Appropriate body substance isolation/personal protective equipment

Equipment that may be helpful:

- Oropharyngeal airways (various sizes)
- Nasopharyngeal airways (various sizes)
- Bag-mask device
- Oxygen cylinder and regulator
- Suction device
- Pulse oximeter
- End-tidal carbon dioxide meter

Indications

- Foreign body airway obstruction in infants younger than 1 year

Contraindications

- Infants who can cry or cough

Complications

- None if properly performed

Procedures

Ensure body substance isolation before beginning procedures.

Prior to beginning patient care, appropriate body substance isolation procedures should be employed.

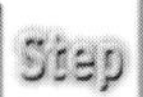 

Assess airway and breathing.

Assess the infant's ability to breathe on his or her own. An infant who can cry or cough is moving air and does not need emergent assistance. Back slaps may be useful to assist the infant to cough.

If it appears that the infant cannot cry or cough, prepare to deliver back slaps and chest thrusts.

Deliver back slaps.

Supporting the head and neck, lift the infant and rotate face down onto your arm with the head lower than the trunk. Support your arm against your thigh if necessary. Deliver *five* upward, glancing back slaps between the shoulder blades using the heel of one hand. This procedure should occur quickly, in less than 3 to 5 seconds.

Deliver chest thrusts.

Supporting the head and neck, rotate the infant face up, keeping the head lower than the trunk. Locate the midsternum by placing two fingers just below the nipple line. Deliver *five* chest thrusts, straight down to a depth one third to one half the depth of the chest. *Do not* use upward, glancing compressions.

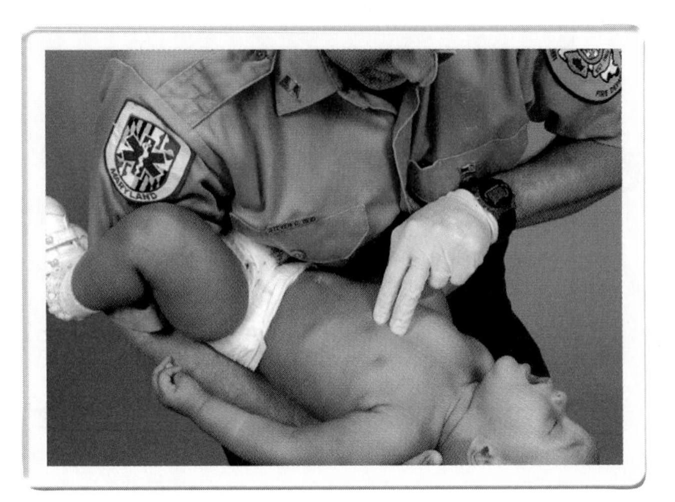

Repeat back slaps and chest thrusts.

Repeat back slaps and chest thrusts until the foreign body is expelled, the infant can breathe spontaneously, or he or she becomes unconscious. If the foreign body does not fall from the mouth, check the mouth to see if it is visible. If it is seen, perform a finger sweep to remove the object. Once the object has been expelled, assess the victim's ability to breathe on his or her own as described in Step 2.

If the infant becomes unconscious, call for help.

Deliver chest compressions (for unconscious patients or for patients who become unconscious while removing an airway obstruction).

Locate the correct position on the chest by placing two fingers just below the midnipple line (see Step 4). Compress the chest vertically one third to one half the depth of the chest. The compressions should be smooth and even, equally up and down in a fluid motion. Deliver the full 30 compressions.

Step 7 Inspect the airway and assess for breathing.

After delivering 30 compressions, open the infant's mouth and look for the dislodged foreign object. If it is seen, perform a finger sweep to remove the object. If the object cannot be seen, do not perform a finger sweep. If no object is seen, attempt to ventilate. If able to ventilate, give two breaths, each lasting 1 full second.

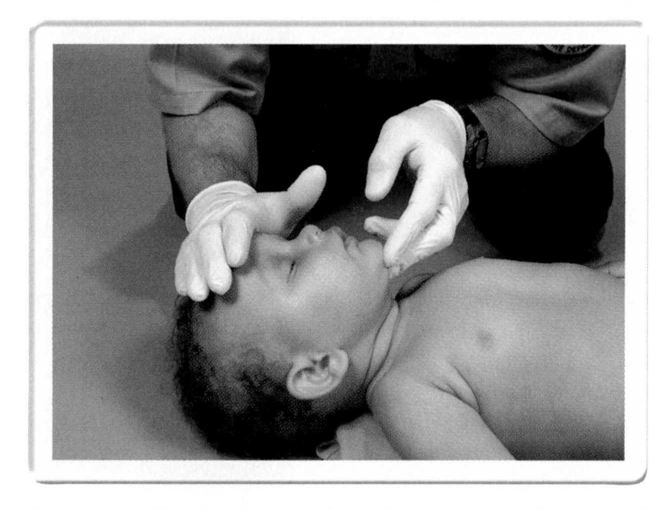

Step 8 Repeat steps 6 and 7 until successful.

Repeat the sequence of 30 compressions and two ventilations until the patient can be adequately ventilated. Remember to inspect the mouth to see if the foreign object can be seen before each set of ventilations. Once the airway has been cleared and the patient can be ventilated, assess the patient to determine the need for infant CPR.

SKILL 17

Neonatal Resuscitation

Performance Objective

Given a neonate patient in distress following delivery, the candidate shall demonstrate the proper sequence and techniques for neonatal resuscitation, within 10 minutes or less.

Equipment

The following equipment is required to perform this skill:

- Appropriate body substance isolation/personal protective equipment

Equipment that may be helpful:

- Oropharyngeal airways (various sizes)
- Nasopharyngeal airways (various sizes)
- Bag-mask device
- Oxygen cylinder and regulator
- Suction device
- Pulse oximeter
- End-tidal carbon dioxide meter

Indications

- Newborn patients not responding to the normal stimuli to breathe or otherwise exhibiting signs of distress
- Cardiac arrest in infants between birth and hospital discharge

Contraindications

- None

Complications

- None if properly applied

Procedures

Ensure body substance isolation before beginning procedures.

Prior to beginning patient care, appropriate body substance isolation procedures should be employed.

Assess history.

Prior to the birth of the infant, perform a history and assessment to determine the likelihood that the baby will need resuscitation. Preterm delivery is the most common risk factor; however, lack of prenatal care or other characteristics that may be known by the mother should also be considered.

Identify the need for resuscitation by assessing four simple characteristics.

1. Was the baby born at or after a full term of gestation?
2. Is the amniotic fluid clear, with no signs of meconium or infection?
3. Is the newborn breathing or crying?
4. Does the baby appear to have good muscle tone?

If the answer to all four of these questions is yes, there is no need for resuscitation. Move to the basic care of the infant, which should include providing warmth, clearing the airway, drying, and assessing the infant's tissue color.

If the answer to any of the above questions is no, proceed through the following steps.

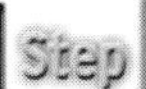

Dry, warm, position, suction, and stimulate the baby and assess further needs.

Upon delivery, wrap the baby in a blanket and cut the cord. Quickly dry the baby of the amniotic fluid and position with the head down. Suction the mouth and nose (if meconium aspiration has occurred).

Stimulate the baby's respirations by gently rubbing the back or flicking the feet with your finger. Assess the need for further interventions by performing an assessment of the baby.

Step 4 Administer oxygen.

Oxygen is needed when the baby fails to breathe adequately or central cyanosis is present. One hundred percent oxygen should be administered using the blow-by technique or a simple neonatal face mask held against the face.

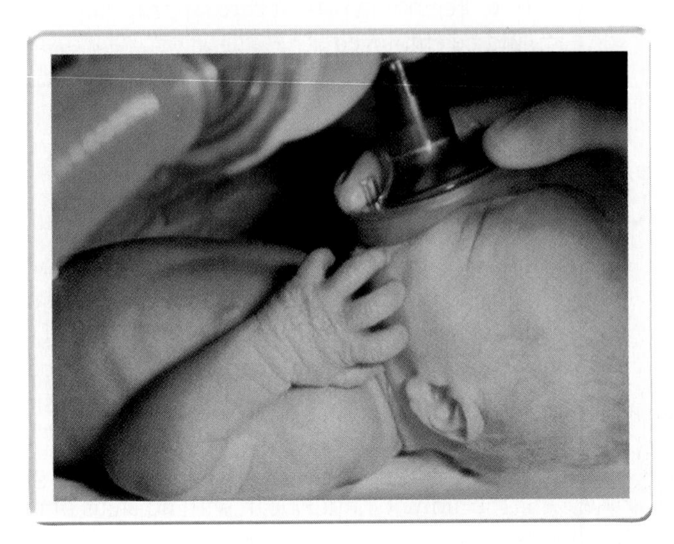

Step 5 Establish effective ventilation.

If the baby is apneic or has gasping respirations, the baby's heart rate is less than 100 beats/min, or central cyanosis persists despite oxygen administration, positive pressure ventilation is indicated.

Ventilations should be performed at a rate of 40 to 60 breaths/min and should be of adequate depth to obtain chest rise. Because neonatal lungs are tight initially, the first few ventilations may require more force than subsequent ventilation.

Step 6 Assess the need for chest compressions.

Closed chest compressions are indicated any time the patient's heart rate is less than 80 beats/min and is not rapidly increasing with ventilations. Compressions should be performed at a rate of 120/min with the sternum depressed one half the depth of the infant's chest, using a 3:1 ratio of compressions to ventilations.

The two-thumb, encircling-hands technique is recommended rather than the standard two-finger technique.

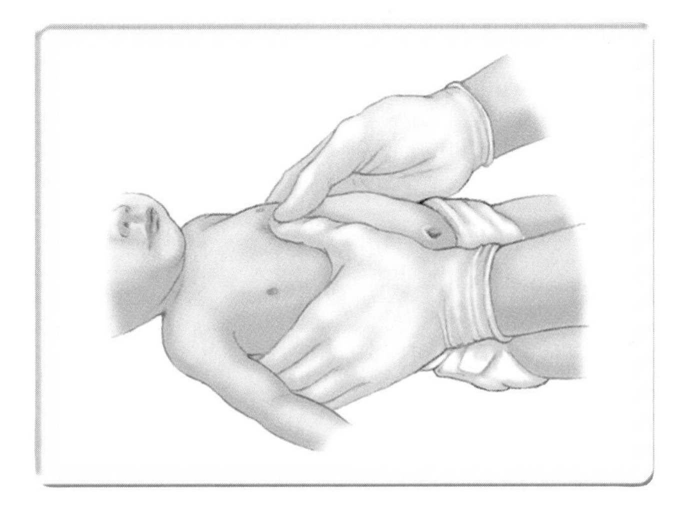

However, the standard two-finger technique may be used for neonatal CPR if preferred.

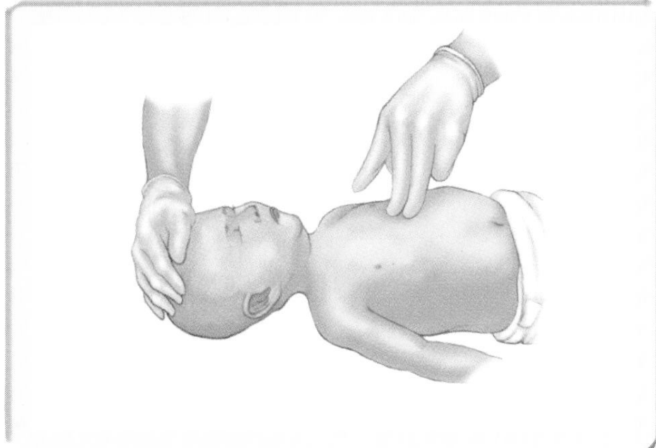

SKILL 18

Cardiac Arrest Management With an AED

Performance Objective

Given the proper equipment, the candidate shall demonstrate proper cardiac arrest management with an automated external defibrillator (AED), using the criteria herein prescribed, in 5 minutes or less.

Equipment

The following equipment is required to perform this skill:

- Appropriate body substance isolation/personal protective equipment
- Automated external defibrillator and pads
- Pocket mask with
  - One-way valve
  - Oxygen connecting port

Equipment that may be helpful:

- Oropharyngeal airways (various sizes)
- Nasopharyngeal airways (various sizes)
- Bag-mask device
- Oxygen cylinder and regulator
- Suction device
- Pulse oximeter
- End-tidal carbon dioxide meter

Indications

- Ventricular fibrillation/nonperfusing ventricular tachycardia

Contraindications

- Infants

Complications

- Asystole

Procedures

Step 1 Ensure body substance isolation before beginning procedures.

Prior to beginning patient care, appropriate body substance isolation procedures should be employed.

Step 2 Briefly question rescuers and witnesses about arrest events.

Before applying the AED, question initial rescuers and witnesses about the events leading to cardiac arrest. Specific questions should include the following:

- What was the time of arrest onset?
- What care was provided before the arrival of rescuers (such as the determination of pulselessness and breathlessness and the initiation of CPR)?
- What was the patient doing before the arrest?
- Did the patient have any complaints before the arrest (such as chest pain or dyspnea)?
- (In some situations) Is there a Do Not Resuscitate order active for this patient?

Step 3 Direct rescuers to stop CPR.

Direct the rescuers performing CPR to stop compressions and ventilations.

Step 4 Verify absence of spontaneous pulse to verify the need to attach defibrillator.

Check the carotid pulse for the presence of a spontaneous pulse for up to 10 seconds. The absence of a pulse is an indication to attach the AED.

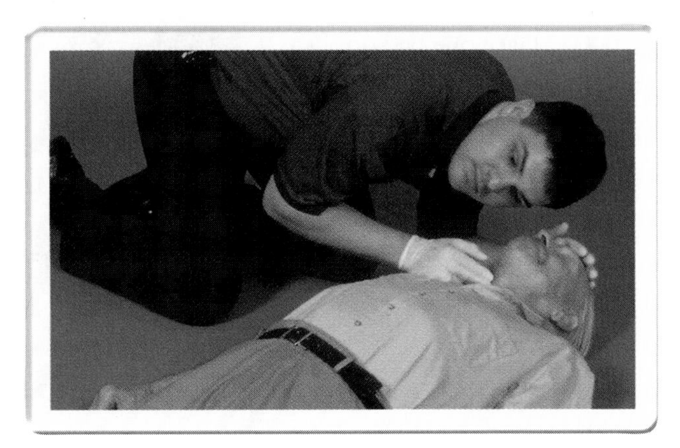

Special Populations

Pediatric AEDs

Newer models of AEDs have been designed with a high specificity in recognizing ventricular fibrillation and ventricular tachycardia in children. Some AEDs also have dose attenuating systems to reduce the energy delivered. These devices are intended for children 1 to 8 years of age. If these devices are available, they should be used when a pediatric patient is in a shockable rhythm. However, if no pediatric-specific AED is available, a standard adult AED should be used in pediatric cardiac arrests.

In the Field

Immediate Defibrillation or Immediate CPR?

When an AED is immediately available, it should be used as soon as possible with a witnessed sudden onset of cardiac arrest. However, when the patient is found in cardiac arrest with a down time reported or assumed to be over 4 minutes, perform CPR for five cycles (2 minutes) of compressions and ventilations before delivering the first shock.

Step 5 Redirect rescuers to continue CPR.

After 10 seconds, if no definitive pulse is found, direct rescuers to continue CPR. Throughout the management of cardiac arrest, interruptions of CPR should be kept to a minimum and should last no longer than is absolutely necessary.

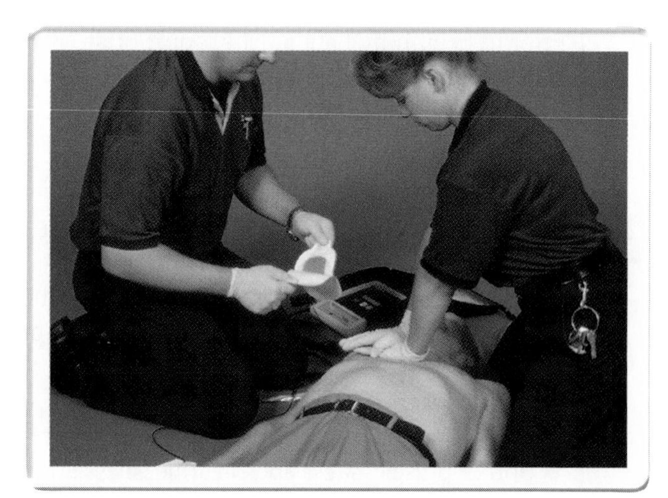

Step 6 Turn on defibrillator power.

Turn on power to the AED and wait for the computer to cycle. Pay attention to any written or verbal messages that occur.

Step 7 Attach automated defibrillator to patient.

Attach the patient pads to the defibrillator cables. If the cables have positive and negative leads, make sure the pads are attached to the correct lead.

Place one pad just to the right of the upper sternum below the right clavicle, and the other pad just below and to the left of the left nipple.

or

Place one pad on the sternum, and the other pad on the patient's back directly posterior to the first pad. The determination of which position to use will be made by either protocol or manufacturer's specifications.

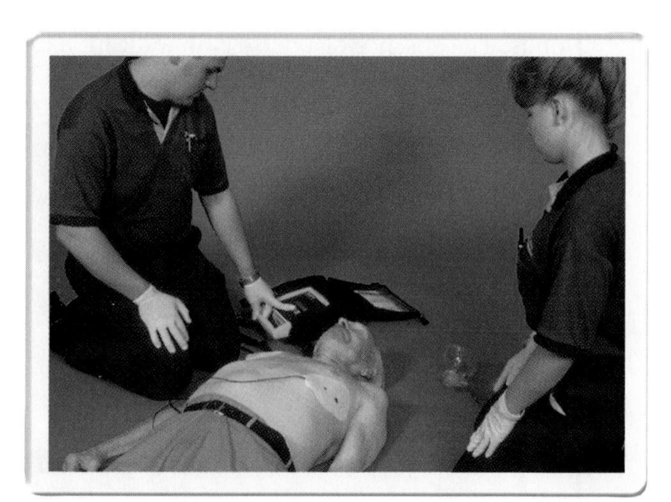

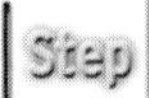 

Ensure all individuals are standing clear of the patient.

Instruct personnel doing CPR to stop CPR after completing their fifth cycle of compressions and ventilations and to stand clear of the patient.

Explain to other rescuers that you are analyzing the rhythm and that patient motion will interfere with this process.

Initiate analysis of rhythm.

Press the Analyze button on the AED and stand clear of the patient, the AED, and the cables. Wait for the AED to analyze the need to defibrillate. All other resuscitative measures should be withheld while this step is in process to avoid delays or false analysis due to patient motion.

Deliver shock.

When advised to defibrillate by the AED, instruct all other rescuers to stand clear. Visually confirm that *all* rescuers and bystanders are clear of the patient, bed, or conductive equipment attached to the patient.

Charge the defibrillator and deliver the shock as directed by the AED.

For monophasic defibrillators, the first and all subsequent shocks should be set at 360 J. For biphasic defibrillators, the energy should be set at the level recommended by the manufacturer. This will usually be between 120 J and 200 J.

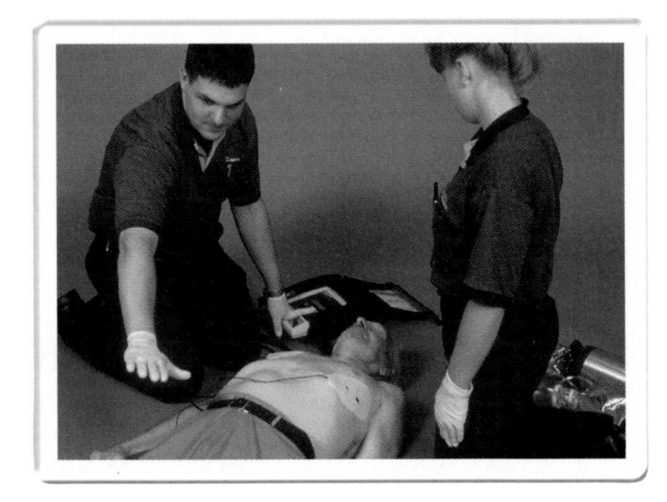

 Step 11

Immediately direct rescuers to continue CPR.

CPR should be started immediately after the shock is delivered, without reassessment of the rhythm or the pulse.

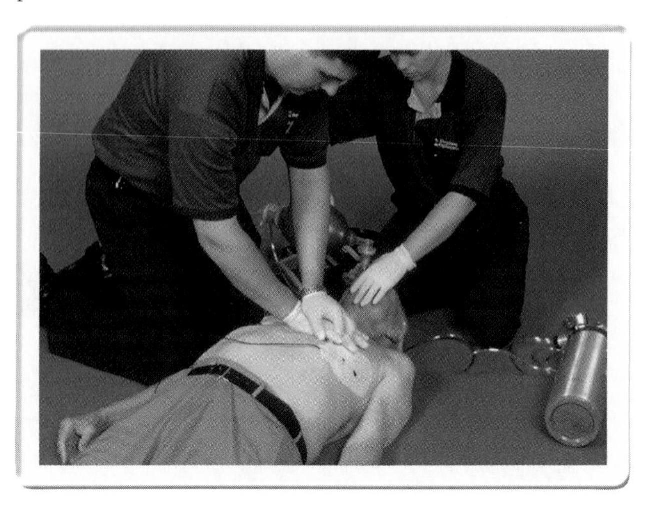

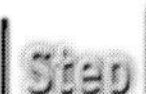  Step 12

Gather additional information on the arrest event.

Question witnesses, especially if they are family or close friends, about the patient's medical history. This should include the following:

- History of cardiovascular disease
- History of respiratory disease
- History of diabetes
- Presence of cardiac risk factors

In some arrest circumstances, especially those involving children and young adults, arrest is not caused by a preexisting cardiac event. Be sure to question witnesses about the possibility of trauma, drug use, and anaphylactic events.

  Step 13

Confirm effectiveness of CPR (ventilation and compressions).

Traditionally, the effectiveness of CPR has been assessed by checking the carotid pulse. Studies have shown that jugular vein reflex waves have been misinterpreted as a pulse, making pulse assessment during chest compressions unreliable. Better means of identifying effective ventilations and compressions are improvement of pulse oximeter readings, improved end-tidal carbon dioxide readings, responsive pupils, and, occasionally, patient movement.

Direct insertion of a simple airway adjunct (oropharyngeal or nasopharyngeal airway).

Direct rescuers to insert an oropharyngeal airway (see Skill 7). Nasopharyngeal airways should be reserved for patients who have been revived and have a gag reflex (see Skill 9).

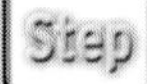

Direct ventilation of patient.

Confirm that ventilations are appropriate and effective. Proper rate and depths should be ensured. If initial ventilations were begun with mouth-to-mask resuscitation, direct rescuers to begin ventilations with a bag-mask device.

Ensure high concentration of oxygen connected to the ventilatory adjunct.

Direct rescuers to begin supplemental oxygen delivery to the bag-mask device. Oxygen should be administered at 15 L/min or greater.

Ensure CPR continues without unnecessary or prolonged interruption.

Ensure that compressions are delivered correctly. Depths and rate of compressions should be monitored. Interruptions should be avoided.

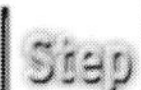

Reevaluate patient after five cycles of CPR.

At the completion of five cycles of 30 compressions and two ventilations, direct rescuers performing CPR to stop compressions and ventilations.

Check carotid pulse for presence of spontaneous pulse. Press the Analyze button on the AED and stand clear of the patient, the AED, and the cables. Wait for the AED to analyze the need to defibrillate. All other resuscitative measures should be withheld while this step is in process to avoid delays or false analysis due to patient motion.

Step 19 Repeat defibrillation sequence.

When advised to defibrillate by the AED, instruct the rescuers to continue CPR. With CPR in progress, charge the defibrillator.

When the AED is charged, instruct the rescuers to stop CPR and stand clear. Once you are sure no rescuers are in contact with the patient, deliver the shock as directed by the AED.

Immediately following the shock, direct the rescuers to continue CPR. After five more cycles of CPR, reassess the patient. Continue the sequence of defibrillation and CPR as needed.

Step 20 Begin transportation of patient.

The patient must be prepared for transport without interrupting CPR unnecessarily. The patient should be placed on a long backboard and secured, and then placed on the cot. Raise the cot only half way to allow for effective chest compression during movement. Carefully and slowly move the patient to the ambulance. Reassessment of the patient should be performed every five cycles of 30 compressions and two ventilations, with defibrillation delivered as indicated.

Check Your Knowledge

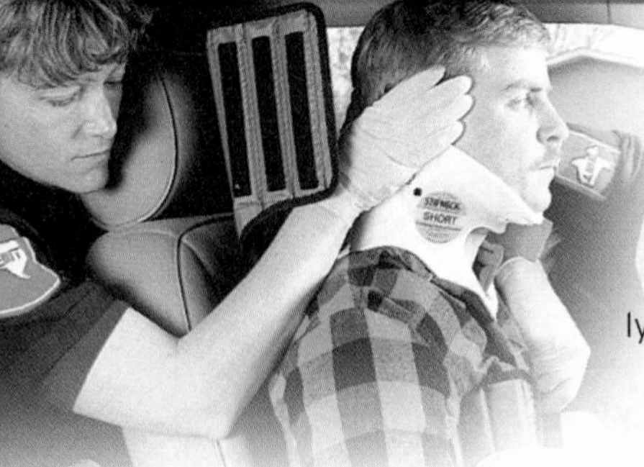

You and your partner are dispatched to a local restaurant for an unresponsive male. You arrive to find a 57-year-old male lying supine on the floor. His wife tells you that he stated he didn't feel well and became very pale and sweaty before losing consciousness. She informs you that she and the waiter lowered the patient to the floor where he is now lying. You determine that the patient is unresponsive and that ALS backup is en route.

1. What is the appropriate method for opening this patient's airway?

A. Head tilt-chin lift maneuver
B. Jaw-thrust maneuver
C. Tongue-jaw lift maneuver
D. None of the above

2. Using the look, listen, and feel procedure, how long should you take to determine breathlessness?

A. < 5 seconds
B. < 10 seconds
C. < 15 seconds
D. < 20 seconds

3. After opening the patient's airway and delivering two slow breaths, you should feel for a __________ pulse for no more than 10 seconds.

A. femoral
B. radial
C. carotid
D. brachial

4. After you determine that the patient is pulseless, you begin chest compressions, compressing the chest:

A. 1 to $1\frac{1}{2}$ inches.
B. $1\frac{1}{2}$ to 2 inches.
C. 2 to $2\frac{1}{2}$ inches.
D. None of the above.

5. After the AED delivers a shock, CPR should be resumed:

A. after 10 seconds.
B. after reassessment for pulse.
C. after the analyze phase of the AED.
D. immediately.

Additional Questions

6. The method for clearing a foreign body airway obstruction in an adult patient who is able to speak is:

A. the Heimlich maneuver.
B. abdominal thrusts.
C. chest compressions.
D. None of the above.

7. The ratio of chest compressions to ventilations in child CPR is:

A. 30:2.
B. 15:2.
C. 15:1.
D. 5:1.

8. When assessing the pulse of an infant, check the ________________ artery.

A. femoral
B. radial
C. carotid
D. brachial

9. The procedure for clearing a foreign body airway obstruction in a conscious infant does NOT include:

A. back blows.
B. chest compressions.
C. looking for visible obstruction.
D. blind finger sweeps.

10. The rate of chest compressions for neonatal resuscitation is:

A. 80/min.
B. 90/min.
C. 100/min.
D. 120/min.

Patient Assessment and Management

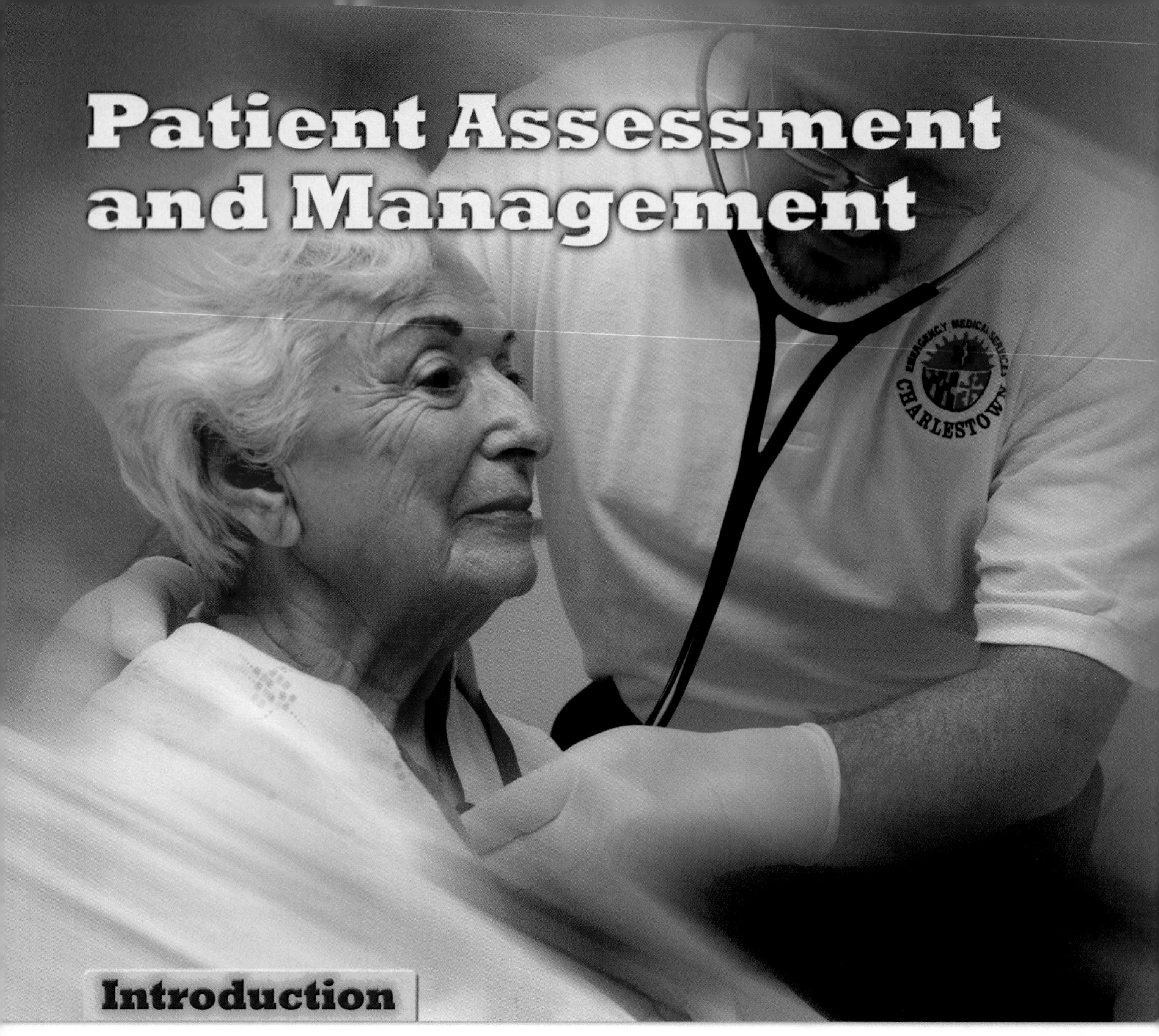

Introduction

Proper assessment of a patient is a necessary prerequisite for proper patient management. Thorough assessments, from the first meeting to transfer of care at the receiving facility, lead to accurate patient care. Oral and written communication skills are vital to this process. For example, the hospital assessment begins with a review of the patient report that is provided by the field personnel.

As you practice patient assessment, consider the required documentation that will finalize the patient care. Good documentation follows good patient assessment; good patient assessment leads to good documentation. The skills in this section will encourage you to master this important process.

SECTION 3

SKILL 19 Patient Assessment and Management

Performance Objective

Given a patient, the proper equipment, and a scenario describing the circumstances of the patient's injury or illness, the candidate shall perform an assessment of the patient, including the scene size-up, initial assessment, focused history and physical examination or detailed physical examination, and state appropriate interventions for the patient's problems, using the criteria herein prescribed, in 15 minutes or less.

Equipment

The following equipment is required to perform this skill:

- Appropriate body substance isolation/personal protective equipment
- Watch with second hand (or digital equivalent)
- Stethoscope with diaphragm and bell
- Sphygmomanometer, with various-sized cuffs
- Pen and paper (patient care chart)

Equipment that may be helpful:

- Pen light
- Thermometer
- Pulse oximeter
- End-tidal carbon dioxide meter

Indications

- Patient care

Contraindications

- None

Complications

- None

Procedures

Ensure body substance isolation before beginning procedures.

Prior to beginning patient care, appropriate body substance isolation procedures should be employed.

▼

Scene size-up: Determine that the scene or situation is safe.

Prior to initiating patient contact, perform a visual sweep of the scene. Look for any hazards to yourself, other rescuers, and bystanders. Hazards should be identified, and information about hazards should be relayed to other responders. *Do not* enter a scene in which hazards are not stabilized.

▼

Determine mechanism of injury or nature of illness, number of patients, and need for additional help.

Assess the mechanism of injury or the nature of illness. Quickly assess the scene to determine whether other patients are present. Do not assume that in medical emergencies no other patients exist. Situations of stress, poisoning, and environmental problems often produce multiple patients.

Request additional resources if the number of patients or the situation exceeds the capabilities of the initial responding unit. This may include calling for additional ambulances, the fire department, heavy rescue, HazMat, or utility companies.

▼

Consider stabilization of the spine (applies when indicated).

Determine whether the patient may have injuries to the spine in association with the chief complaint. If present, apply appropriate stabilization procedures.

▼

Step 5 Initial assessment: Form general impression of patient.

Gather a general impression of the patient upon initial contact. Determine whether the patient looks stable or unstable from his or her initial appearance. Be prepared to change your mind as the survey continues and improvement or deterioration is found. In conscious patients, a simple battery of questions can determine the chief complaint and provide detailed information about the mechanism of injury or nature of illness.

Step 6 Determine responsiveness or level of consciousness.

Quickly determine the patient's level of consciousness through use of the AVPU scale (**A**wake, responds to **V**erbal stimuli, responds to physical or **P**ainful stimuli, **U**nconscious).

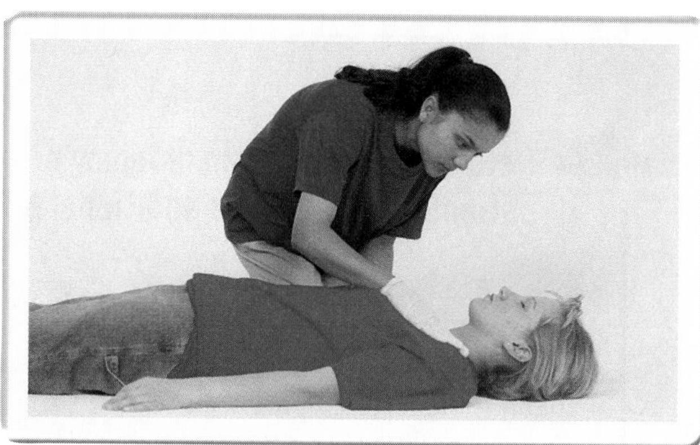

Step 7 Assess airway (with consideration of cervical spine).

Patients who are talking have an open airway. Quickly evaluate the patency to ensure the airway remains open throughout patient care.

In the unresponsive patient, open and assess the airway. Determine the possibility of cervical injury before opening the airway. If cervical injury is suspected, open the airway using the jaw-thrust maneuver (see Skill 1). Consider inserting an oropharyngeal or nasopharyngeal airway if indicated (see Skill 7 or Skill 9). Quickly inspect the neck for injuries that might interfere with ventilation.

Step 8 Assess breathing.

Assess breathing and ensure adequate ventilation.

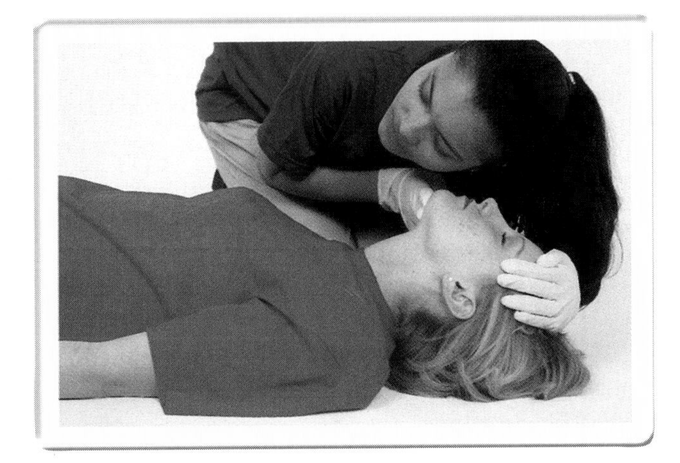

Assess the patient's chest for injuries that might compromise ventilations (tension pneumothorax, flail chest, or sucking chest wound).

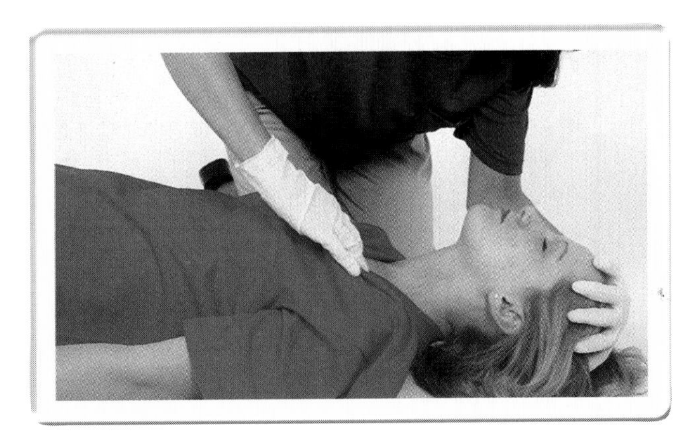

If the ventilations are inadequate, initiate bag-mask ventilations. Apply supplemental oxygen as indicated.

Step 9 Assess circulation.

Assess for and control major bleeding. Assess the patient's pulse for rate and strength. The patient's skin and tissues should be assessed for tissue color, temperature, and condition. Initiate shock management, including keeping the patient warm and elevating the legs if indicated.

Step 10 Determine chief complaint and apparent life threats.

From the information gathered, determine the patient's chief complaint and identify apparent threats to life.

Step 11 Identify priority patients and make transport decision.

In most cases, the goal in patient care is to begin transport as soon as practical—usually within 10 minutes for trauma patients, and within 15 minutes for medical patients. Although trauma patients *cannot* be stabilized in the field and require rapid transport, many medical patients can be stabilized in the field. The opportunity to transport should not impede the appropriate delivery of care. Always ask the question, Will the patient benefit more from rapid transport or from field treatment?

From the information so far obtained, determine the patient's priority of care and determine the need for immediate or delayed transport. This can be one of the most difficult decisions associated with patient care. Always go back to the basics: ensure an airway, maintain breathing, and maintain circulation. If these steps cannot be performed, and all available methods available have been employed, transport without delay.

Priority and transport decisions are not one and the same. A patient who is unconscious has, by definition, an unstable airway. Thus, the patient is a high priority. However, the placement of an oropharyngeal airway or endotracheal tube stabilizes the patient, lowering the priority and making transport less urgent. On a similar note, a conscious and alert patient who is developing severe and worsening chest discomfort and dyspnea and has not responded to any therapy you have attempted is a high-priority patient who needs rapid transport. You should note that high priority is always rapid transport, whereas low priority is always delayed transport.

High-priority patients. Altered level of consciousness, airway or ventilatory compromise, poor systemic circulation (low systolic blood pressure), multisystem trauma, etc.
Low-priority patients. Isolated injuries, minor bleeding, minor medical complaint, stable vital signs within normal limits, etc.
Rapid transport. Uncontrolled airway, difficulty ventilating, signs and symptoms of shock, multisystem trauma, compromised medical condition, etc.
Delayed transport. Minor injuries, minor medical complaint, etc.

In the Field

Cardiac Arrest

- **For the EMT.** Cardiac arrest is a rapid, ***controlled*** transport. The patient needs effective CPR, which cannot be performed during transport.
- **For the Paramedic.** Cardiac arrest is ***not*** a rapid transport situation. The patient will benefit more from proper care delivered *where the patient was found* than from hurried care and rapid transport.

Step 12 Determine depth of focused history and physical examination.

- **Trauma patients with no significant mechanism of injury** *or* **responsive medical patients.** Focus on the body systems involved and the specific injury or illness, guided by the patient's chief complaint (see Steps 13 through 17 and Steps 29 through 32).
- **Trauma patients with serious injuries or mechanisms of injury** *or* **unresponsive medical patients.** Perform a rapid head-to-toe physical examination (DCAP-BTLS). Assess the size, equality, and reactivity of the pupils; examine the neck for distended jugular veins and tracheal deviation; examine the chest for crepitation, paradoxical motion, and equality of breath sounds; and examine the abdomen for rigidity or distention (see Steps 18 through 32).

Trauma Patients With No Significant Mechanism of Injury or Responsive Medical Patients

Step 13 Obtain baseline vital signs (see Skill 20).

Determine the patient's level of consciousness (using the patient's orientation to person, place, and time), pulse (rate, strength, and regularity), respirations (rate, depth, and quality), blood pressure, skin temperature and condition, and tissue color.

Step 14 Patient history: Gather history of present illness and/or complete a SAMPLE history.

Gather the history of present illness from the patient, or from family or bystanders if necessary.

- **Cardiac or respiratory patients.** Follow the standard OPQRST approach.
  - Onset of symptoms: Sudden or gradual.
  - Provocation: What worsens or lessens the condition.
  - Quality: Sharp, throbbing, crushing, dull, etc.
  - Radiation: Whether the pain or condition travels to other parts of the body.
  - Severity: Describe using a scale of 1 to 10, with 10 being the worst pain ever felt by the patient.
  - Time: How long ago this condition began.
- **Altered mental status patients.** Determine a description of the episode, the onset, the duration, and associated evidence of trauma. Question the patient or family about any interventions already taken, and the possibility of seizures or fever.
- **Allergic reaction patients.** Determine history of allergies, what the patient was exposed to and how, the effects and progression of exposure, and any interventions already performed (such as the use of antihistamines).

continued

- **Poisoning/overdose patients.** Determine the toxic substance, the time at which exposure occurred, the type of exposure (ingestion or contact), how much toxin was involved, and over how long a period of time. Also determine what interventions were performed. The patient's weight and the effects of exposure are also important.
- **Obstetrical patients.** First determine the possibility of pregnancy ("Are you pregnant?" is a good start). Determine the length of pregnancy (in weeks, if known), EDC (estimated date of confinement), any pain or contractions, bleeding or discharge, and possible crowning. Does the patient feel a need to push? Determine the patient's last menstrual period, previous pregnancies, and the results of each pregnancy (gravida, para, ab, c-sections)
- **Behavioral patients.** Start by asking "How do you feel?" Determine thoughts of suicide and whether the patient is a threat to self or others. Ask about any underlying medical problem, and any interventions already performed.
- **Environmental emergency patients.** Determine the cause of the problem (heat, cold, water, or altitude), the length of exposure, and any loss of consciousness. Determine what effects have occurred and whether the effects are general or local.

Gather the SAMPLE history from the patient, or from family or bystanders if necessary.

Symptoms (history of present illness)
Allergies to medications
Medications currently taken by the patient (prescription and nonprescription)
Past history—pertinent medical information related to current condition
Last oral intake (food, drink), or last menstrual period in obstetrical/gynecological patients
Events leading up to emergency

Apply appropriate interventions.

Provide interventions appropriate for the patient's condition.

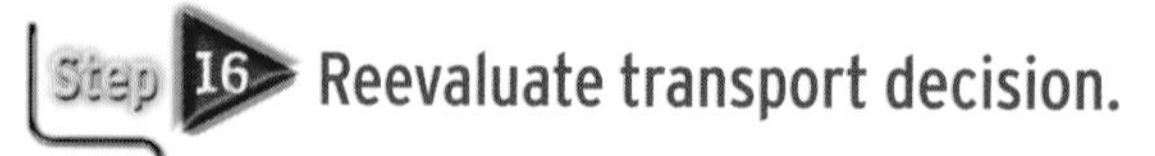

Reevaluate transport decision.

Reevaluate the initial decision to transport. Based on the total picture now available, determine whether to initiate transport or to slow the emergency response.

Step 17 Consider completing a detailed physical examination.

If indicated, perform a detailed physical examination, including the following:

- **Head.** Inspect/palpate the scalp; inspect/palpate the face; inspect nose, mouth, and ears; inspect eyes and pupils. Look for bleeding, bruising, discoloration, and deformity.

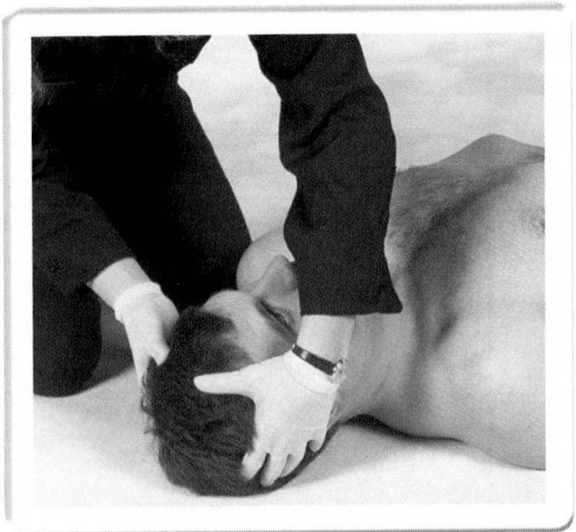

- **Neck.** Inspect/palpate the neck; look for jugular vein distention and tracheal deviation.

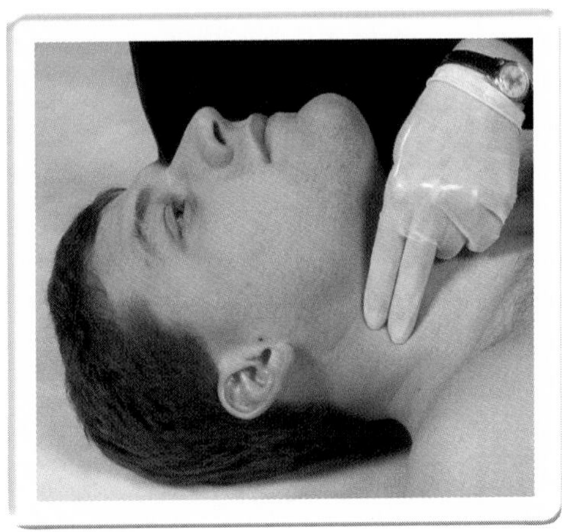

- **Chest.** Inspect/palpate the stability of the chest wall; auscultate the equality of breath sounds.

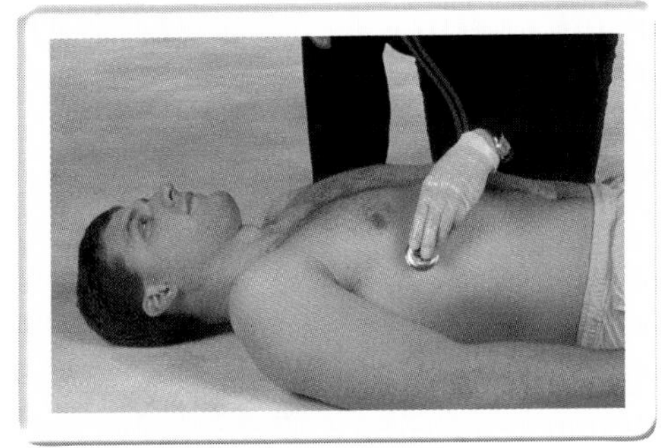

Step 17 continued

- **Abdomen.** Inspect/palpate all four quadrants. Look for bruising and distention.

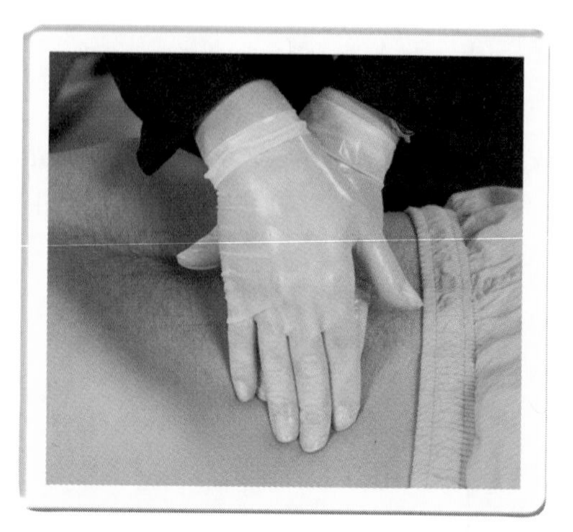

- **Pelvis.** Palpate for instability. Inspect genitalia and perineum, as appropriate.

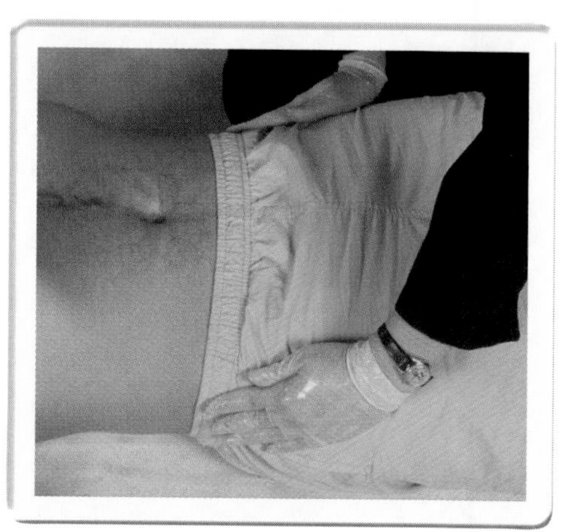

- **Extremities.** Inspect/palpate all four extremities. Look for quality of circulation and motor and sensory function.

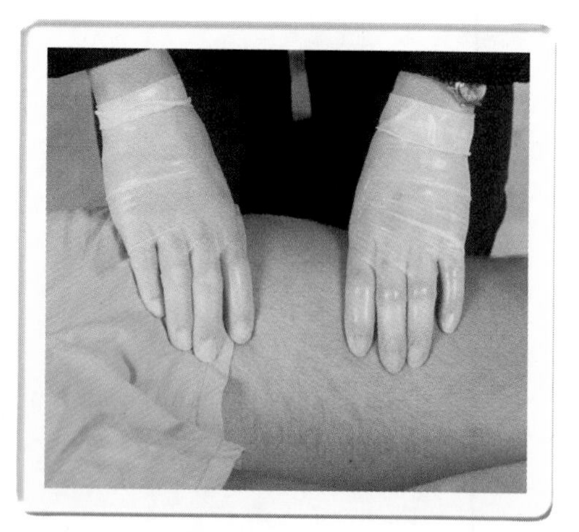

- **Back.** Inspect/palpate the thoracic/lumbar spine and the entire back. Make this a thorough inspection. If the patient will be placed on a backboard, this will be the last exam of the back you will be able to perform (move to Step 29).

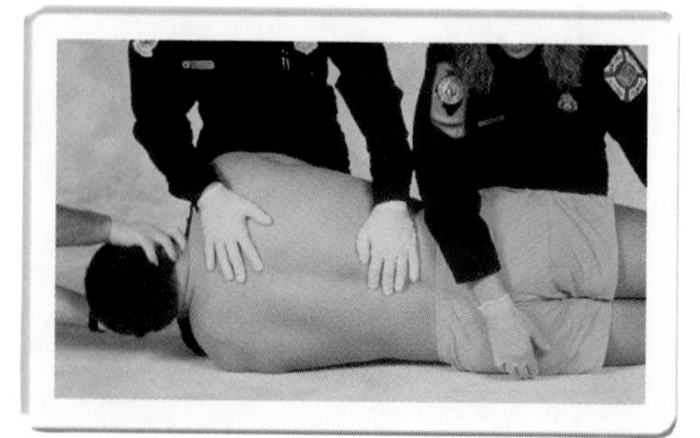

Trauma Patients With Significant Mechanisms of Injury or Unresponsive Medical Patients

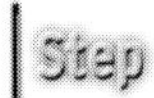

Step 18 Obtain or direct assistant to obtain baseline vital signs (see Skill 20).

Determine the patient's level of consciousness (using the patient's orientation to person, place, and time), pulse (rate, strength, and regularity), respirations (rate, depth, and quality), blood pressure, skin temperature and condition, and tissue color. Perform a rapid physical exam as described in Steps 19 through 26.

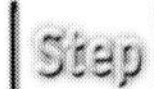

Step 19 Assess the head.

Inspect/palpate the scalp; inspect/palpate the face; inspect nose, mouth, and ears; inspect eyes and pupils. Look for bleeding, bruising, discoloration, and deformity.

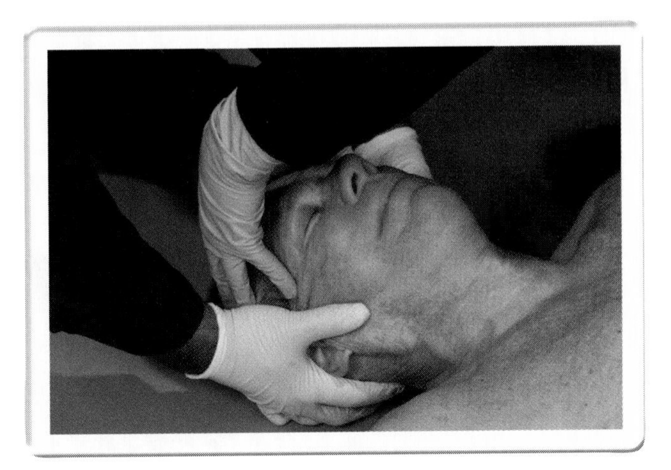

Step 20 Assess the neck.

Inspect/palpate the neck; look for jugular vein distention and tracheal deviation. Apply cervical spine immobilization device on trauma patients.

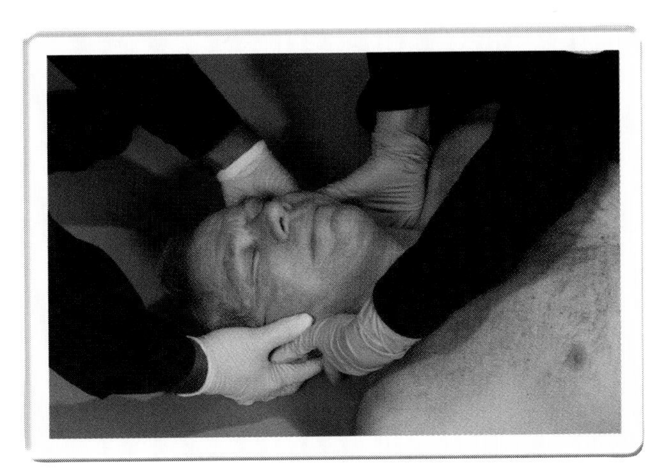

Assess the chest.

Inspect/palpate the stability of the chest wall; auscultate the equality of breath sounds.

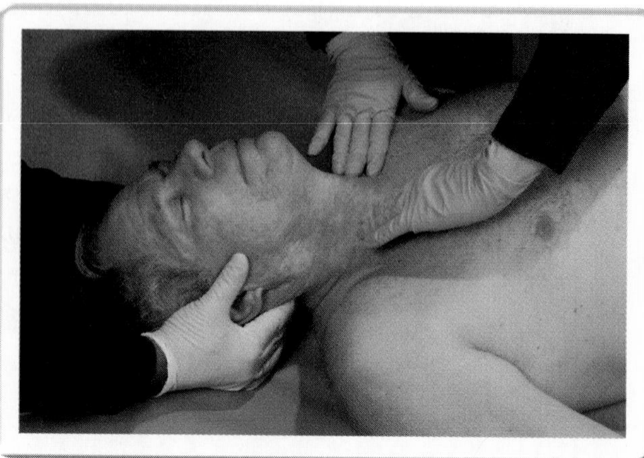

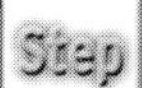

Assess the abdomen and pelvis.

Inspect/palpate all four quadrants. Look for bruising and distention. Palpate for instability in the pelvis. Assessment of genitalia and perineum should be performed if appropriate.

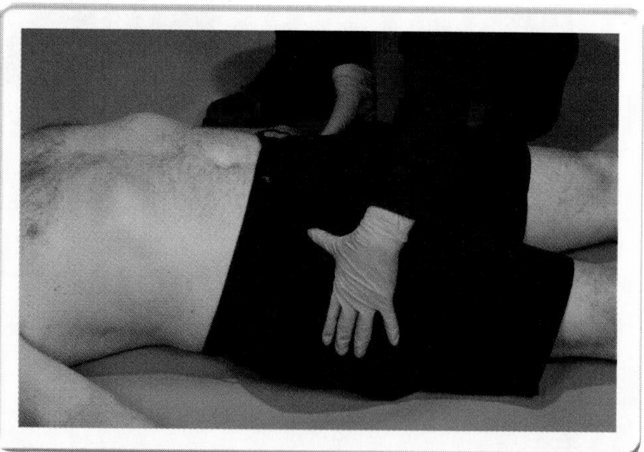

Step 23 Assess lower extremities.

Inspect/palpate the lower extremities. Look for quality of circulation and motor and sensory function. Assess the patient's ability to move joints, and to move the foot against resistance.

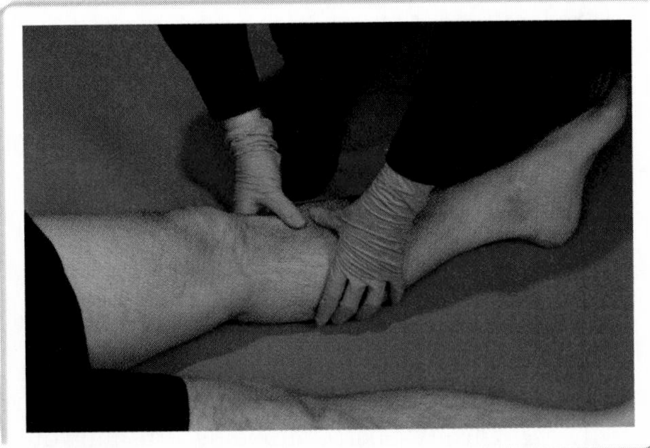

Step 24 Assess upper extremities.

Inspect/palpate the upper extremities looking for the quality of circulation and motor and sensory function. Assess the patient's ability to move joints. Assess grip strength in both hands.

Step 25 Assess the back.

Inspect/palpate the thoracic/lumbar spine and the entire back. Make this a thorough inspection. If the patient will be placed on a backboard, this will be the last exam of the back you will be able to perform.

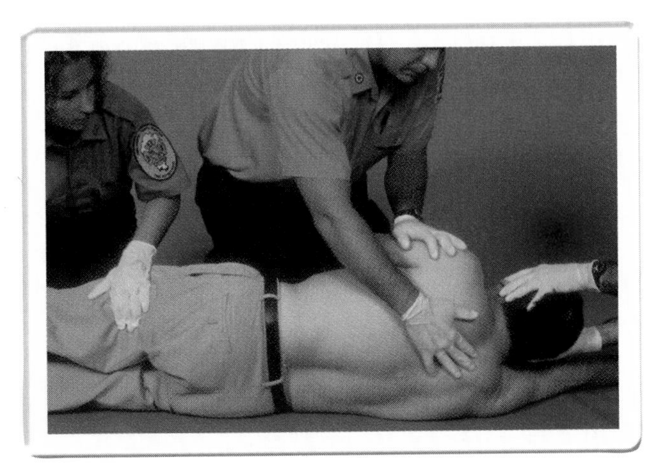

Step 26 Manage secondary injuries and wounds appropriately.

Once initial threats to life have been managed, secondary injuries should be managed if the situation allows. This includes stabilizing minor fractures and minor bleeding (move to Step 27).

Step 27 Patient history: Gather history of present illness.

Gather the history of present illness from the patient, or from family or bystanders if necessary.

- **Cardiac or respiratory patients.** Follow the standard OPQRST approach.
  - Onset of symptoms: Sudden or gradual.
  - Provocation: What worsens or lessens the condition.
  - Quality: Sharp, throbbing, crushing, dull, etc.
  - Radiation: Whether the pain or condition travels to other parts of the body.
  - Severity: Describe using a scale of 1 to 10, with 10 being the worst pain ever felt by the patient.
  - Time: How long ago this condition began.
- **Altered mental status patients.** Determine a description of the episode, the onset, the duration, and associated evidence of trauma. Question the patient or family about any interventions already taken, and the possibility of seizures or fever.
- **Allergic reaction patients.** Determine history of allergies, what the patient was exposed to and how, the effects and progression of exposure, and any interventions already performed (such as the use of antihistamines).
- **Poisoning/overdose patients.** Determine the toxic substance, the time at which exposure occurred, the type of exposure (ingestion or contact), how much toxin was involved, and over how long a period of time. Also determine what interventions were performed. The patient's weight and the effects of exposure are also important.
- **Obstetrical patients.** First determine the possibility of pregnancy ("Are you pregnant?" is a good start). Determine the length of pregnancy (in weeks, if known), EDC (estimated date of confinement), any pain or contractions, bleeding or discharge, and possible crowning. Does the patient feel a need to push? Determine the patient's last menstrual period, previous pregnancies, and the results of each pregnancy (gravida, para, ab, c-sections)
- **Behavioral patients.** Start by asking "How do you feel?" Determine thoughts of suicide and whether the patient is a threat to self or others. Ask about any underlying medical problem, and any interventions already performed.
- **Environmental emergency patients.** Determine the cause of the problem (heat, cold, water, or altitude), the length of exposure, and any loss of consciousness. Determine what effects have occurred and whether the effects are general or local.

Step 28 Complete the SAMPLE history.

Gather the SAMPLE history from the patient, or from family or bystanders if necessary.

Symptoms (history of present illness)
Allergies to medications
Medications currently taken by the patient (prescription and nonprescription)
Past history—pertinent medical information related to current condition
Last oral intake (food, drink), or last menstrual period in obstetrical/gynecological patients
Events leading up to emergency

Step 29 Ongoing assessment: Report to hospital after obtaining appropriate information.

The purpose of the hospital radio report is to alert the emergency department (ED) of the arrival of a patient. In most cases the information should be as brief as possible. Only in cases where specific treatment orders are desired should a detailed report be given. The information relayed will be used by the emergency department staff to ensure the proper triage of the patient upon arrival.

The report should follow an organized format, giving in order the following information:

- Identification of unit and self
- Age and gender of patient
- Problems found in primary assessment, if any
- Patient's chief complaint
- Estimate of severity of patient's condition
- Brief history of present illness or injury
- Vital signs
- Pertinent treatment provided
- Anticipated time of arrival

This is Medic 417, EMT Williams. We are en route to your facility with a 21-year-old female involved in a motor vehicle collision. The patient is stable, conscious, and complaining of shoulder and chest pain with dyspnea. Her pulse is 100 beats/min and regular, blood pressure 114/76 mm Hg, and respirations are 18 breaths/min and regular. We should be in the ED in 15 minutes.

Note: Upon arrival at the receiving facility, a more detailed patient report should be given. This detailed report should be a complete description of the patient's presentation, care delivered, and response to interventions.

Step 30 Repeat initial assessment and vital signs.

Repeat the initial assessment. Vital signs should be repeated every 5 minutes for unstable patients (those in the high-priority or rapid-transport categories) and every 15 minutes for stable patients. Reassess the patient's chief complaints and repeat the focused physical exam as necessary. Determine the effects of the interventions performed. A minimum of three sets of recorded vital signs should have been performed upon arrival at the hospital.

Step 31 Evaluate response to interventions.

Following interventions, evaluate the patient to determine if the interventions were effective. Look for improvement in respiratory effort and perfusion, as well as deterioration in condition. Also look for adverse effects from interventions.

Step 32 Repeat focused assessment regarding patient complaint or injuries.

If indicated, repeat the focused assessment. This will be essential if responses to interventions are not obvious.

Vital Signs

Performance Objective

Given a human patient, the candidate shall assess and interpret a patient's pulse, respirations, palpated and auscultated blood pressure, pulse oximetry, and end-tidal carbon dioxide level, in 6 minutes or less.

Equipment

The following equipment is required to perform this skill:

- Appropriate body substance isolation/personal protective equipment
- Watch with second hand (or digital equivalent)
- Stethoscope with diaphragm and bell
- Sphygmomanometer, with various-sized cuffs
- Pen and paper (patient care chart)
- Pulse oximeter
- End-tidal carbon dioxide meter

Equipment that may be helpful:

- Reflex hammer
- Pen light
- Thermometer

Indications

- General patient assessment
- Determination of patient status

Contraindications

- Blood pressure should not be assessed on the ipsilateral arm of a radical mastectomy patient or on patients with a hemodialysis fistula or an indwelling peripherally inserted central catheter.
- Some patients may have specific reasons why blood pressure should be taken on one arm or another. Those wishes should be honored.

Complications

- None

Procedures

Step 1 Ensure body substance isolation before beginning procedures.

Prior to beginning patient care, appropriate body substance isolation procedures should be employed.

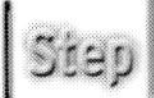

Step 2 Calculate radial pulse.

Using the first two fingers, palpate and count the rate of the radial pulse. Describe the quality of the pulse (strong, weak, thready, etc.), and the regularity of the rhythm. Interpret the findings.

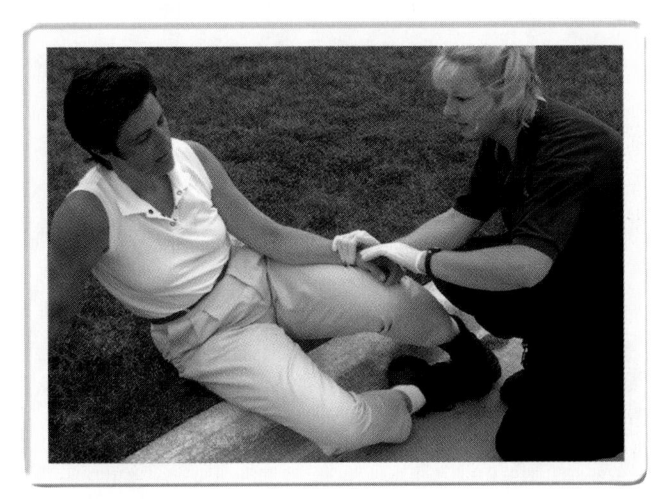

Step 3 Calculate respiratory rate.

Count the rate of the patient's respirations and describe quality (shallow, labored, etc.) and regularity of the respiratory pattern. Interpret the findings.

Step 4 Palpate systolic blood pressure.

Apply the cuff snugly to the upper arm, at least 1 inch above the antecubital fossa.

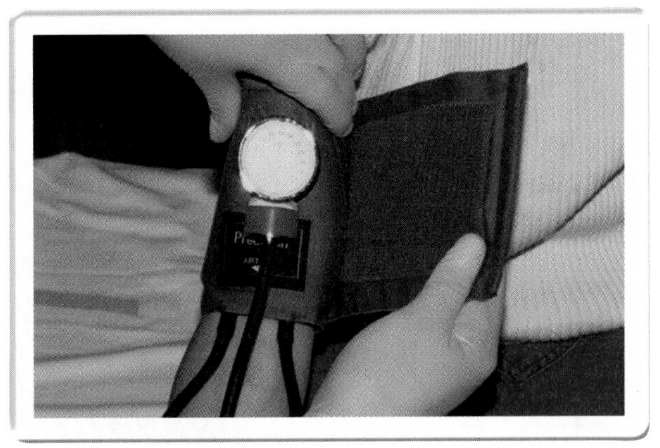

Locate the radial pulse using two fingers. Keep the fingers in place throughout this procedure.

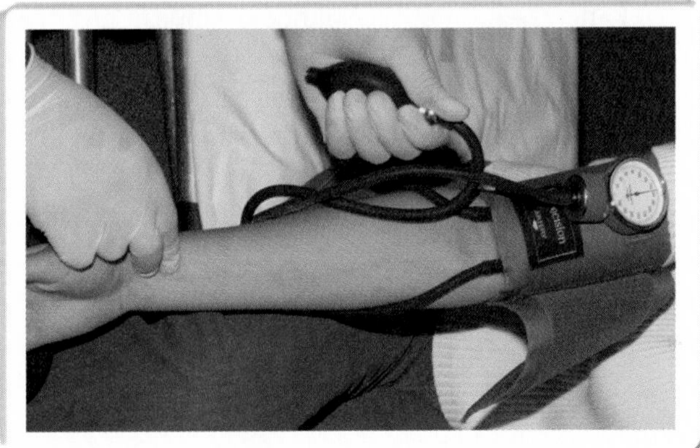

Inflate the cuff gently, feeling for the disappearance of the palpable pulse. After the pulse disappears, give the bulb one or two more squeezes. Carefully and slowly release the air from the blood pressure cuff. Palpate for the return of the pulse. This is the palpated systolic pressure. Deflate the cuff completely when finished and prior to reinflation (if a second or third attempt is performed). Interpret the findings.

In the Field

Sizing the Blood Pressure Cuff

Choosing the proper blood pressure cuff for the patient is an important part of obtaining an accurate blood pressure.

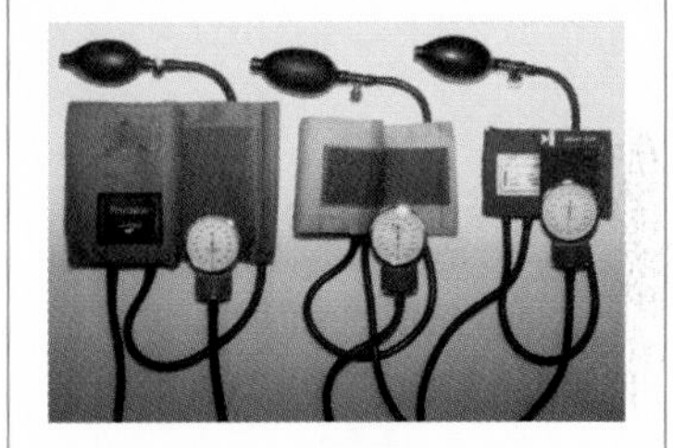

Most blood pressure cuffs have markings to identify the correct size. When the cuff is placed around the patient's arm, the arrow or other markings should fit between the minimum and maximum size indicators.

Step 5 Auscultate blood pressure.

Apply the cuff snugly to the upper arm, at least 1 inch above the antecubital fossa. Place the diaphragm of the stethoscope over the brachial artery.

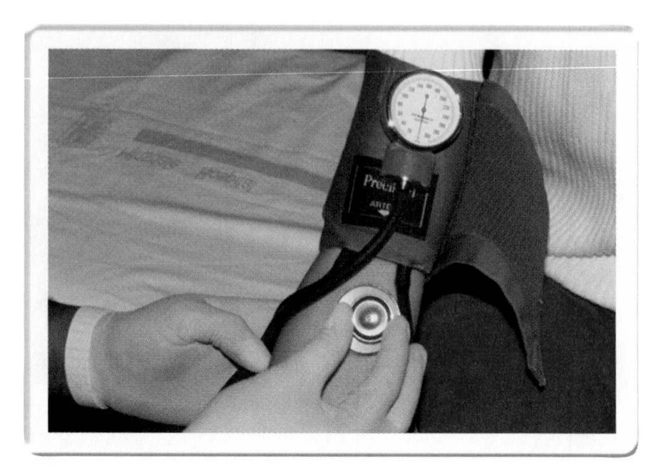

Inflate the cuff gently, listening for Korotkoff sounds to appear as the pressure increases. After the sounds disappear, give the bulb one or two more squeezes. Carefully and slowly release the air from the blood pressure cuff. Listen for the return and disappearance of the Korotkoff sounds. The return of Korotkoff sounds gives the systolic blood pressure, whereas the disappearance gives the diastolic. Deflate the cuff completely when finished and prior to reinflation (if a second or third attempt is performed). Interpret the findings.

Advanced Vital Signs

 6 **Interpret pulse oximetry.**

Attach the pulse oximeter sensor to the patient's finger. Correctly read the display and interpret the findings.

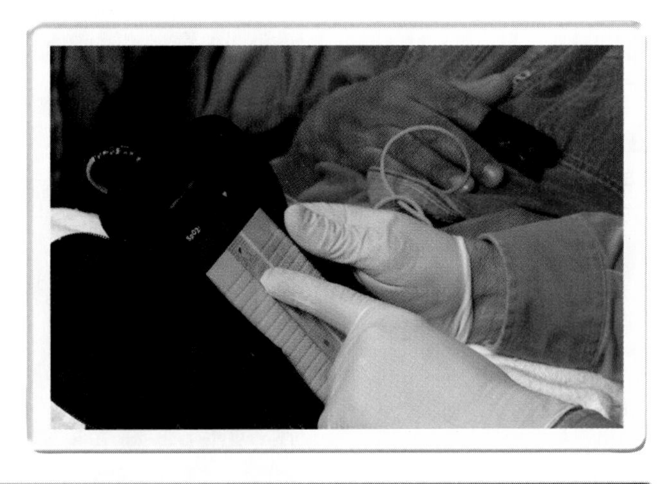

  **Interpret capnography.**

Attach the capnography cannula to the patient. Correctly read the numeric and wave form display and interpret the findings.

SKILL 21

Communication

Performance Objective

Given all required information concerning the assessment and management of a patient, the candidate will relay necessary information to receiving personnel, in 2 minutes or less.

Equipment

The following equipment is required to perform this skill:

- Appropriate body substance isolation/personal protective equipment
- Voice skills
- Radio

Indications

- Radio reports for the purpose of:
  - Notifying the receiving facility of the status of the incoming patient
  - Consulting with the receiving physician concerning further patient care
- Face-to-face report for the transfer of care

Contraindications

- Relaying of patient information to personnel not involved in the continuation of care is prohibited under the Health Insurance Portability and Accountability Act (HIPAA), and may violate other privacy laws as well.

Complications

- None

Procedures

Radio Report for Notification

The purpose of the hospital radio report is to alert the emergency department (ED) of the arrival of a patient. In most cases the information should be as brief as possible. The information relayed will be used by the ED staff to ensure the proper triage of the patient upon arrival.

Step 1 Key microphone.

Key the microphone and wait 1 second before talking.

Step 2 Contact receiving facility.

With the microphone held approximately 1″ from your lips, make initial contact with the receiving facility. It is best to identify the receiving facility first, and then your unit. Stating the hospital name first "wakes up" the receiver and keeps them alert to who is calling.

> Memorial Emergency, this is Medic 16.

Step 3 Give patient report.

After the receiving facility acknowledges the radio call, key the microphone and wait 1 second before talking. Then, with the microphone held approximately 1″ from your lips, begin the notification report.

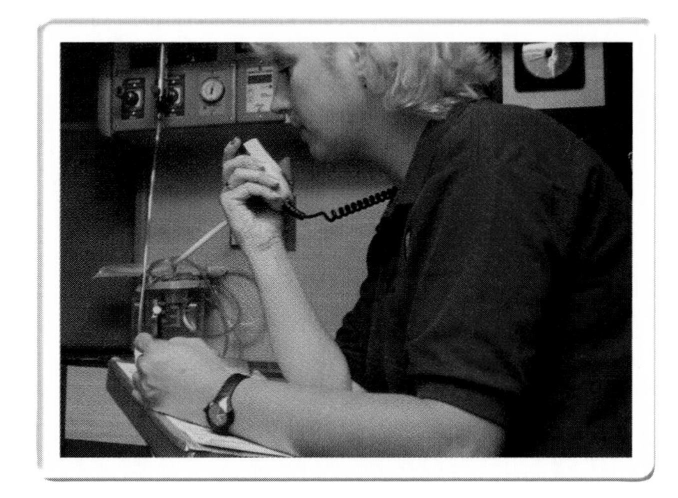

continued

The report should follow an organized format, giving in order the following information:

- Age and gender of patient
- Problems found in primary assessment, if any
- Patient's chief complaint
- Estimate of severity of patient's condition
- Brief history of present illness or injury
- Vital signs
- Pertinent treatment provided
- Anticipated time of arrival

This is Medic 16, paramedic Jackson. We are en route to your facility with a 21-year-old female involved in a motor vehicle collision. The patient is stable, conscious, and complaining of shoulder and chest pain with dyspnea. Her pulse is 100 beats/min and regular, blood pressure is 114/76 mm Hg, and respirations are 18 breaths/min, regular and nonlabored. We should be in the ED in 15 minutes.

Radio Report for Physician Consultation

The purpose of the physician consultation is to obtain a second opinion on the care of the patient, to determine whether the care of the patient needs to follow a path not described in protocol, or to discuss alternative regimens not in the standing orders. Although brief is always better, achieving proper consultation requires that the physician be given enough information to form a complete picture of the patient's situation.

Key microphone.

Key the microphone and wait 1 second before talking.

Contact receiving facility.

With the microphone held approximately 1″ from your lips, make initial contact with the receiving facility and ask for physician consultation. It is best to identify the receiving facility first, and then your unit. Stating the hospital's name first "wakes up" the receiver and keeps them alert to who is calling.

> Memorial Emergency, this is Medic 25. Need to speak with a physician for consultation please.

Give detailed patient report.

After the physician acknowledges the radio call, key the microphone and wait 1 second. Then, with the microphone held approximately 1″ from your lips, begin the consultation report. The report should follow an organized format, giving in order the following information:

- Identification of unit and self
- Age and gender of patient
- Problems found in primary assessment, if any
- Patient's chief complaint, including OPQRST
- History of present illness or injury
- Current findings: Vital signs, presentation, ECG, etc.
- Care rendered
- Description of request (orders needed)

continued

This is Medic 25. We are on the scene with a 58-year-old male who is complaining of retrosternal chest pain radiating into his jaw, left shoulder, and arm. He states that the pain began while watching TV and has been growing steadily worse for the past 2 hours. He denies any previous history of heart problems. At this time he is conscious and alert, denying nausea or shortness of breath. He is showing a sinus rhythm with frequent multiformed premature ventricular contractions on the monitor. Breath sounds are clear and equal bilaterally. Blood pressure is 104/68 mm Hg, pulse 76 beats/min and irregular, respirations 28 breaths/min, skin cool and clammy. We have placed him on 4 liters of oxygen and started an IV of normal saline to keep vein open. The patient has shown no improvement. After 2 mg of morphine, the patient is still in considerable pain. There is no change in the respiratory status or blood pressure. I would like to give additional morphine, up to 10 mg.

Face-to-Face Transfer of Care

The purpose of the face-to-face report is to provide the receiving personnel with all of the necessary information about the patient's previous and current condition and the care rendered by EMS.

Step 1 Introduce patient to receiving personnel.

Introduce the patient by name to the receiving personnel who will be assuming care. In cases of an unresponsive patient, introduce the receiving personnel to the patient and tell the patient what is happening. Never assume an unresponsive patient is not aware of what is going on around him or her.

Mr. Smith, this is Sylvia. She is the nurse who will be taking care of you here in the emergency department.

Step 2 Deliver detailed patient report.

Proceeding in an organized fashion, give a detailed description of the patient's condition, giving in order the following information:

- The patient's age and history of chief complaint
- Chief complaint and associated symptoms
- Original assessment findings
- Care given
- Current assessment findings
- Known allergies and medications
- Pertinent past medical history
- Status of continuing interventions (endotracheal tube placement, IV fluid infusion, next required medication doses, drip medications, etc.)

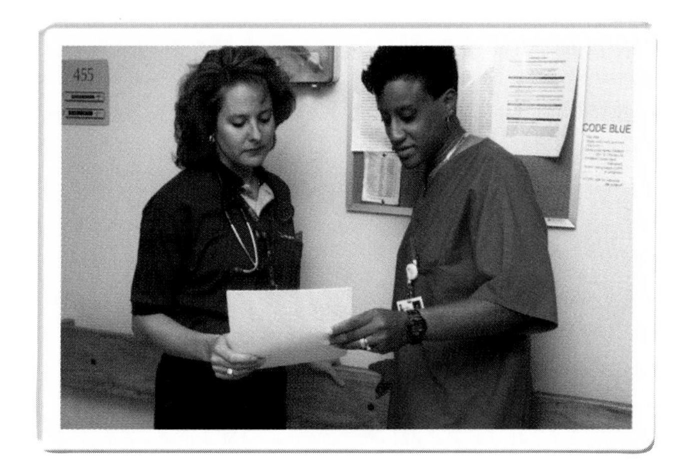

continued

Mr. Smith is 62 years old and woke up this morning complaining of a little nausea. After about 3 hours of taking antacids he decided he needed to call the ambulance. When we arrived at his house he was sitting in the living room in moderate distress. He admitted to slight chest pain with the nausea and numbness in his left arm. He rated the pain as a 6. No significant dyspnea reported. We did a 12-lead and found some elevation in the inferior leads, but did not see any changes on the right side. His original vitals were all in normal ranges. We placed him on 4 liters of oxygen, gave him one 325-mg aspirin, and started the IV, a 16 gauge there in the left forearm. He then got one nitro spray and 4 mg of morphine. Currently he still has slight pain rated at a 3, ECG is an ectopy-free normal sinus at 94 beats/min, respirations 14 breaths/min, and blood pressure 128/86 mm Hg. The IV is still running with normal saline to keep vein open; he's had about 50 mL so far, and there are no signs of infiltration or infection.

Documentation

Performance Objective

Following the care of a patient, the candidate will demonstrate proper documentation of a complete patient report, in 30 minutes or less.

Equipment

The following equipment is required to perform this skill:

- Appropriate body substance isolation/personal protective equipment
- Pen
- Patient care report

Equipment that may be helpful:

- Dictionary
- Thesaurus
- Approved abbreviation list

Indications

- Terminal component of every patient contact, whether patient was transported or not

Contraindications

- None

Complications

- Omitted information can lead to mistakes in the postarrival care of the patient.
- Poorly written reports cast doubt on the quality of patient care.
- Delays in writing reports lead to omitted information and a decline in detail.

Procedures

Writing the SOAP Patient Report

SOAP is an acronym for Subjective, Objective, Assessment, and Patient care, the four components of the patient report. Using this method, specific pieces of information regarding the patient examination and history are divided into two groups: those that cannot be determined, witnessed, or proven by the examiner (subjective) and those that can be determined, witnessed, or proven by the examiner (objective). In other words, this report is a simple division of signs from symptoms.

Step 1 Subjective

Subjective information includes the patient history, whether or not the history comes from the patient directly. Bystander or witness accounts and information supplied by family or neighbors are included in this section, as are the patient's age, past medical condition, allergies, medication taken, chief complaint, and associated complaints, and the onset and duration of the problem.

All subjective information is taken at the patient's word rather than being determined by the EMT. Pain, nausea, dyspnea, and dizziness are also subjective findings because none can be proven to exist by the EMT.

Also included under subjective information are denials from the patient or bystanders, such as loss of consciousness.

Step 2 Objective

Objective information can be determined by the examiner. The location and position of the patient, as well as his or her surroundings, skin temperature, pulse, respiratory rate, and blood pressure, are all objective findings. These findings cannot be denied or exaggerated by the patient.

The objective findings should be limited to the *initial* physical examination only, and point the assessment toward a specific plan of treatment. All findings from the initial physical examination, vital signs, oxygen saturation (SaO_2), end-tidal carbon dioxide level, and blood glucose level are recorded in this section.

Therapy that was initiated prior to the arrival of EMS should also be documented in this section, such as cardiopulmonary resuscitation performed by a bystander, an IV initiated by a first responder, or bleeding control applied through first aid.

Step 3 Assessment

The assessment is a process that helps clarify the patient's problem as seen by the examiner. It is important to note that although an examiner cannot make an official medical diagnosis, an examiner's diagnosis can and must be formed. The patient is treated based on this diagnosis. Without a diagnosis of some kind, treatment of the patient cannot begin.

The examiner's determination of the patient's condition or problem should match the signs and symptoms noted in the subjective and objective areas of the report, and should point toward a specific protocol for treatment. Although the examiner's diagnosis or opinion of the patient's condition does not need to be as explicit as the physician's official diagnosis, it should reflect the basis for any treatment performed, based on written protocols.

Patient care (plan)

The patient care section of the report is a narrative description of the care provided to the patient. The term *plan*, used on physician charts, is not appropriate for a prehospital patient report. When physicians write a plan, they are writing orders to nursing staff and laboratory personnel describing procedures to be carried out in the patient care. Since the "plan" of action on an EMT's report is actually a retrospective of treatment that has already occurred, a more narrative approach to the actual patient care rendered should be taken.

The patient care section should begin with a description of where the patient encounter initiated and note any additional persons who were important in the assessment of the patient. This should be followed by the assessment of the airway, breathing, and circulation and what determination was made about their status. The report should then proceed to describe each step in the assessment and management of the patient.

This section should also include the response to *all* interventions, any improvement or deterioration in the patient's condition, and what therapeutic adjustments were performed during the course of treatment. The closing of the patient care section and the report should include the transfer of care at the receiving destination. This will include documentation that a report was given to a nurse or physician, as well as the status of continued therapies.

Section	Purpose	Content
Subjective	History and symptoms	What you can't prove, can't see, can't feel
Objective	Physical exam and signs	What you found, saw, heard, felt, smelt
Assessment	Examiner's diagnosis	Determined and supported by the subjective and objective section findings
Patient care (plan)	Narrative postscript	Detailed explanation of how the assessment, patient interview, therapies, and interventions were carried out; must include responses to all interventions and final disposition of the patient

Patient Report Quality Assurance Checklist

This checklist is to be used to ensure that each clinical patient report contains the necessary information. This is a BLS checklist; EMT-Intermediates and paramedics should use the advanced checklist.

Subjective Findings

Are all of the following clearly identified as appropriate?

__ 1 Chief complaint
__ 2 Associated complaints (positive or negative)

SAMPLE History

__ 3 **S**ymptoms (Positive or negative)
__ 4 **A**llergies: Drug and food
__ 5 **M**edications: Prescription, over the counter, nonprescription, herbal, compliance
__ 6 **P**ast medical/surgical history
__ 7 **L**ast oral intake
__ 8 **E**vents preceding or leading up to the injury or illness

OPQRST of Chief Complaint

__ 9 Onset
__ 10 Provocation
__ 11 Quality
__ 12 Radiation
__ 13 Severity
__ 14 Time
__ 15 Social history: Smoking, alcohol use, drug use, etc.
__ 16 Caller: Patient, family, bystanders, etc.
__ 17 Source of information: Patient, family, bystanders, etc.

Special Considerations

Allergies/Anaphylaxis

__ 18 Allergies
__ 19 Specifics of previous reactions

Environmental

__ 20 Exposure
__ 21 Thermal protection
__ 22 Fluid/electrolyte intake

Gynecologic/Obstetric

__ 23 Last menstrual cycle
__ 24 Possibility of pregnancy
__ 25 Due date
__ 26 Gravida, para, ab
__ 27 Bleeding/discharge

Poisoning/Overdose

__ 28 Toxin (alone or with alcohol)
__ 29 How much
__ 30 Over how long

Trauma: Motor Vehicle

__ 31 Restraints used
__ 32 Patient location in vehicle

Denials

__ 33 Head: Loss of consciousness, headache, dizziness, trauma in the last 3 weeks
__ 34 Eyes: Visual difficulty, photophobia, discharge, pain
__ 35 Ears: Change or ringing, discharge
__ 36 Nose: Congestion, pain, discharge
__ 37 Mouth/throat: Pain, trauma or lesions, difficulty speaking/swallowing
__ 38 Chest: Dyspnea, chest pain, palpitations, cough
__ 39 Abdomen: Pain, nausea, emesis, diarrhea
__ 40 GU: Dysuria, hematuria, polyuria, incontinence, discharge
__ 41 Neck/back/extremities: Pain, numbness, tingling
__ 42 Skin: Rashes, lesions, itching

Objective Findings

Are all of the following documented as appropriate for this patient?

Scene Size-up

__ 43 Scene description (mechanism of injury, environment, vehicle damage, etc.)
__ 44 Patient description and position (weight, supine, prone, driver, back-seat passenger, etc.)

Initial Assessment

__ 45 Level of consciousness (AVPU or estimated Glasgow)
__ 46 Airway
__ 47 Breathing (air movement and effort)
__ 48 Circulation
__ 49 General impression

Physical Examination

Note positive findings for DCAP-BTLS in the following areas. Positive or negative findings should be noted for all other areas as appropriate.

Head

__ 50 Specific examination: Ecchymosis (mastoid, periorbital), drainage (nose, ears), tissue color (conjunctiva, gums), nasal flaring, eye movement, pupil size and response, sclera

Neck

__ 51 Specific examination: Jugular vein distention, carotid bruits, tracheal palpation, spinal palpation, subcutaneous emphysema

Chest

__ 52 Specific examination: Chest rise, paradoxical motion, sucking wounds, breath sounds, respiratory patterns, speech dyspnea, heart sounds, retractions

Abdomen

__ 53 Specific examination: Rigidity, guarding, masses, pulsations, bowel sounds, ecchymosis, rebound, fetal heart tones

Pelvis/Perineum

__ 54 Specific examination: Stability, bleeding, crowning, discharge, incontinence

Back (if not included elsewhere)

__ 55 Specific examination

Extremities

__ 56 Specific examination: Motion, sensation, distal circulation, capillary refill, reflexes, strength

Skin

__ 57 Specific examination: Temperature, moisture, tissue color

Trauma Summation

__ 58 "No other deformities, contusions, abrasions, punctures, burns, lacerations, or swelling noted."

Vital Signs and Technology

__ 59 Respiratory rate and quality
__ 60 Pulse rate, strength, and regularity
__ 61 Blood pressure (orthostatic)
__ 62 Electrocardiogram (ECG)
__ 63 12-lead
__ 64 Oxygen saturation (Sao_2); room air and on oxygen
__ 65 End-tidal carbon dioxide level
__ 66 Blood glucose
__ 67 Core temperature
__ 68 I-STAT lab findings
__ 69 Glasgow Coma Scale score
__ 70 Trauma score (on trauma)

Assessment

Is this assessment supported by the signs and symptoms listed in the Subjective and Objective sections?

__ 71 Matches subjective and objective
__ 72 No other problems identified
__ 73 Written as medical diagnosis

Patient Care

Describe in narrative format from patient contact to transfer of care, with times as appropriate.

__ 74 Initial contact
__ 75 Assessment of ABCs
__ 76 History and assessment
__ 77 Initial interventions
__ 78 Response to interventions

__ 79 Reassessment
__ 80 Additional interventions
__ 81 Response to interventions
__ 82 Placement and position on cot
__ 83 Movement to ambulance
__ 84 Loading and securing
__ 85 Movement to final destination
__ 86 Patient report and transfer of care (staff name), and condition of patient
__ 87 Critical status at transfer

Procedures

Intravenous Catheterization

__ 88 Size, gauge, and type
__ 89 Location
__ 90 Fluid type and rate
__ 91 Success vs number of attempts–provider

Medication Administration

__ 92 Drug, dose, route, time
__ 93 Intramuscular, subcutaneous: Absence of blood return
__ 94 Intravenous: Positive blood return
__ 95 Provider

Endotracheal Intubation

__ 96 Size and depth
__ 97 Confirmation: visualization, breath sounds, condensation, end-tidal carbon dioxide, esophageal detector (at least three)
__ 98 Secured and collar
__ 99 Reconfirmation with each move: lifting, defibrillation, loading onto cot, loading into ambulance, unloading (constant waveform capnography acceptable)
__ 100 Success vs number of attempts–provider

ET Suctioning

__ 101 Return

Chest Compressions

__ 102 Rate, depth, ratio, pulse generation

Ventilations

__ 103 Rate, volume, compliance

Spinal Motion Restriction

__ 104 Method and procedures (log roll, slide, etc.)
__ 105 Circulation, sensation, and motor function before and after procedure

Orthopaedic Immobilization

__ 106 Method and procedures
__ 107 Circulation, sensation, and motor function before and after procedure
__ 108 Anatomic position
__ 109 Elevated

Defibrillation/Cardioversion

__ 110 Joules and response

Pacing

__ 111 Rate and amperage
__ 112 Capture (Electrical/mechanical)

Special Circumstances

Patient Refusal

__113 Informed of need for care and/or transportation
__114 Informed of consequences of not accepting care or transportation
__115 Competency to refuse treatment or transport
__116 Advised of critical signs and symptoms

Treat and Release

__117 Exam performed
__118 Justification of decision why transport is not indicated
__119 Accompanied by (mother, spouse, etc.)
__120 Advised to seek medical care within __ hours or days

Deceased

__121 Obvious signs of death: Rigor mortis, dependent lividity, decapitation, etc.
__122 Time
__123 DNR order

Crime Scene

__124 Police incident report number
__125 Disposition of clothing
__126 Preservation of evidence: Bagging of hands, etc.

Family Violence

__127 Document location given of nearest family violence center
__128 Document injuries that may have resulted from family violence

Child Abuse/Elder Abuse

__129 Police incident number (confirming notification)

Termination of Effort

__ 130 Medical control approval
__ 131 ECG/end-tidal carbon dioxide at time of termination and no change for 10 minutes

Physician Consultation

__ 132 Physician consultation occurred, including name and time

In the Field

Documentation Methods

In the prehospital setting, several methods of documentation are used and are acceptable:

- **Narrative method.** Probably the simplest method to write, but the most difficult to decipher. This method consists of documenting the patient encounter from start to finish using standard prose. The strength is that it is simple to learn. The weakness is that important details are often omitted.
- **CHART or CHARTE method.** CHART stands for Chief complaint, History and physical exam, Assessment, treatment (Rx), and Transport. To use CHARTE, add Exceptions. The strength is that it breaks the care and treatment down into smaller sections. This makes it easier to locate specific assessments or care without reading the entire report. The weakness is that it is difficult to learn.
- **SOAP method.** This is one of the more common forms of documentation and is gaining popularity. It is simple to learn and, once completed, provides a simple means for the reader to review the assessment and management.

SOAP Example 1

Subjective

Arrived to find a 55-year-old male complaining of *chest pain (1)*. He admits to *slight dyspnea as well, with mild nausea (2)*. He describes the chest pain as *extreme pressure (11)*, *retrosternal (3), radiating to the left shoulder and jaw (12)*. *According to his wife (17), he has had angina for the past 3 years (6)* and takes *nitroglycerin and nifedipine (5)*. The patient also admits to a history of *hypertension (6)* and *smokes a pack a day (15)*. The *pain is rated as an "8"* on a scale of 1 to 10 (13). He states the pain began *suddenly (9), while sitting watching the news (8)*, and has been *present for the past 2 hours (14). Sitting quietly relieves the pain slightly, but not to any significant amount (10)*. The patient *denies dizziness, palpitations, or a history of recent cough (38)*. The patient also *denies allergies to drugs or food (4)*. His *last oral intake was dinner just before the pain began, approximately 2 hours ago (7)*.

Objective

The patient was found *sitting on his couch in the living room (44)*. He is *approximately 110 kg (44), diaphoretic (49, 57)*, and *breathes with some distress (47)*. He is *alert; oriented to person, place, and time (45)*; and has a *strong, regular radial pulse (48)*. On physical examination the patient is found to have *ashen oral mucosa (50), equal and reactive pupils approximately 4 mm in diameter (50)*, and *nondistended jugular veins (51)*. His *breathing is slightly labored with equal chest rise (52), breaths sounds are clear and equal bilaterally (52)*, and *no heart murmurs or third heart sounds are detected (52). No edema is noted in the sacral area (55) or ankles (56). Distal circulation, sensation, and motor function are found in all extremities (56)*. His *skin is cool to the touch (57)*, his *respirations are 16 breaths/min and slightly labored (59), pulse is 108 beats/min, strong, and regular (60)*, and *blood pressure 136/94 mm Hg (61)*. His *oxygen saturation is 96% on room air (64)*. The *Glasgow Coma Scale score is 15 (69)* and *blood glucose is 140 mg/dL (66)*.

Assessment

Acute myocardial infarction (71, 73). Cardiac arrest, postresuscitation coma (71, 73).

Patient Care

Initial examination performed in the *patient's living room, with him remaining in his chair (74)*. The *patient's ABCs were assessed and determined adequate (75)*. The *history was gathered from the patient and his wife (76)* while *vital signs were gathered by assisting personnel (76)*. The patient verbally consented to care and transportation. *Oxygen was administered by nasal cannula at 4 L/min (77)*. The *noninvasive blood pressure was applied (77)*. One *325-mg aspirin tablet was administered by mouth at 19:04 with 1 ounce of water (92)*. The patient was *assisted to the cot and positioned in a semi-Fowler's position and secured with straps (82)*. The patient was then *moved to the ambulance (83), loaded, and secured (84). Transport was begun to the hospital (85)*. A firefighter accompanied the EMT in the ambulance. At 19:11, one minute into the transport, the *patient looked up and said he felt "strange" (78). He then lost consciousness (78). The head of the cot was lowered to place the patient in a supine position (80). Assessment found the patient without a pulse or respirations (79). The AED was attached; it advised to shock (80). A single shock was delivered at 150 J (110). Immediate chest compressions were started at a rate of 100 per minute (102). After 30 compressions, ventilations were given by bag-mask, creating chest rise (80). CPR was continued for 2 minutes before reassessment of the patient found a carotid and weak radial pulse (102). No immediate return of consciousness occurred, and breathing remained absent (79). Ventilations continued with a bag-mask at 10 breaths/min, creating chest rise (80). Oxygen was attached to the bag in less than 1 minute (80). The noninvasive blood pressure showed a blood pressure of 103/68 mm Hg, and the oxygen saturation was 98% (79, 81). No spontaneous ventilations or gag reflex were present (79). Upon arrival at the hospital the patient was taken to trauma room 1, where he was transferred onto the hospital cot (86). The report was given to the RN and the doctor (86). The noninvasive blood pressure and oxygen were changed over to the hospital's systems (86). At the time of transfer his blood pressure was 112/74 mm Hg, pulse 94 beats/min, and respirations assisted at 10 breaths/min (86). The end-tidal carbon dioxide was 36 mm Hg with appropriate waveforms (87). No spontaneous respirations were present (87).*

SOAP Example 2

Subjective

The patient is a 27-year-old male *involved in an industrial accident (8)*. He is *complaining of severe pain to his entire body (1)* following a *flash explosion (9)*. He says his *throat is burning and he is having trouble breathing (1)*. His *coworker (17)* says they were *working on a gas water heater when a spark caused the tank to explode (8)*. The *patient was immediately engulfed in flames, and other workers doused the flames with a garden hose (8)*. The patient says he is very healthy with *no past medical history (6)*, takes no *prescription medications (5)*, and is *allergic to sulfa drugs (4)*. His *last oral intake was 3 hours earlier, when he had a full breakfast (7)*. At the time of the explosion the *patient was wearing a polyester/cotton golf shirt, jeans, and boots (9)*. The coworkers state the *shirt was pretty much burned off the patient (17)*. Currently the pain is severe, rated as a *"10" on a scale of 1 to 10 (13)*, and he says *he is very cold (2)*. The incident occurred approximately *10 minutes prior to the arrival of EMS (14)*.

Objective

The *approximately 80 kg patient (44)* was *found lying in a puddle of water with parts of an exploded water tank all around him (43). Coworkers were at his side dousing the patient with cool running water (43)*. The patient is *awake (45)*, shivering, and *moaning in pain (47)*. His *voice is raspy and he is struggling to breathe (47)*. A *weak radial pulse is present (48)*. He is *obviously in distress from severe burns with respiratory compromise (49)*. On physical examination the patient has *second- and third-degree burns to his face and back of head, circumferential neck and chest, anterior abdomen, and both arms (50, 51, 52, 53, 56, 57)*: an estimated *54% body surface area (57). Soot and charring are present on the patient's teeth, and the oral cavity has mucosal burns (50). Eyelashes and eyebrows are singed and barely perceptible (50). Pupils are equal and reactive to light at 3 mm (50)*. There is *slight clouding of the right eye (50). Stridor is audible without a stethoscope (52). Equal bilateral chest rise is present with clear breath sounds (52), heart sounds reveal no murmurs or third heart sounds (52)*. The *shirt is burned into the chest in several areas (52)*. The patient has *palpable pulses, sensation, and movement in all extremities (56). Abrasions and small lacerations are seen over the face and both arms (50, 56). No other deformities, contusions, abrasions, punctures, lacerations or swelling are noted (58)*. He has *respirations of 22 breaths/min, labored (59); pulse is 102 beats/min, strong and regular (60)*; and *blood pressure is 134/84 mm Hg (61)*. His *Spo_2 is 98% on room air (64), blood glucose 140 mg/dL (66)*, and his *core temperature is 96.2 degrees (67)*. His initial *Glasgow Coma Scale score is 15 (69)*, and his *trauma score is 12 (70)*.

Assessment

Critical thermal burns with respiratory compromise (71, 73)

Patient Care

Initial assessment and management was conducted at scene (74). The *airway was assessed and determined to be in danger (75); an air transport company was called to transport the patient to the burn unit at the hospital (77)*. Consent for treatment and transport was received from the patient. *He was removed from the puddle of water (77)* and *placed on a dry sheet on the cot and covered with a second dry sheet and blanket (82, 77). Oxygen, 12 L/min by nonrebreathing mask, was applied at 09:42 (77)*. The patient immediately *began to relax and said he felt warmer and more comfortable (78)*. His *pain was not eased (78)*.

Upon the arrival of air transport, the patient was *transferred to the helicopter stretcher (85)*. The patient was *taken to the helicopter and secured head aft (85)*.

Check Your Knowledge

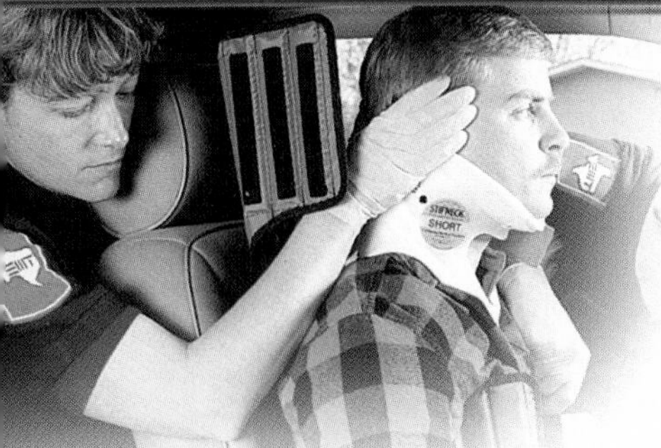

You are dispatched to a local construction site for a fall victim. You arrive to find a 25-year-old male lying on the ground near a ladder. Bystanders tell you that he fell from near the top of what appears to be an approximately 30′ ladder.

1. After determining that the scene is safe, your next step should be to:
- **A.** form a general impression of the patient.
- **B.** determine responsiveness or level of consciousness.
- **C.** assess the airway.
- **D.** determine the mechanism of injury or nature of illness, number of patients, and need for additional help.

2. Initial assessment of this patient should include:
- **A.** the ABCs.
- **B.** determining the responsiveness and level of consciousness.
- **C.** determining the chief complaint.
- **D.** all of the above.

3. A complete set of baseline vital signs includes all of the following EXCEPT:
- **A.** blood pressure.
- **B.** pulse.
- **C.** skin color.
- **D.** grip strength.

4. The last area of the body you will assess in the detailed physical examination is the:
- **A.** head.
- **B.** chest.
- **C.** back.
- **D.** extremities.

5. When assessing the abdomen and pelvis, you must remember to check for:
- **A.** bruising.
- **B.** distention.
- **C.** instability.
- **D.** all of the above.

6. You must assess the patient's pulse for rate, rhythm, and ______________.
- **A.** time
- **B.** quality
- **C.** rebound
- **D.** pace

7. Blood pressure should not be assessed on the ipsilateral arm of a patient who has/had:
- **A.** a radical mastectomy.
- **B.** a hemodialysis fistula.
- **C.** an indwelling peripherally inserted central catheter.
- **D.** all of the above.

8. Your radio report to the receiving facility should include all of the following EXCEPT:
- **A.** patient's chief complaint.
- **B.** anticipated time of arrival.
- **C.** patient's name.
- **D.** vital signs.

9. You will document your initial physical examination findings in which section of the written patient report?
- **A.** Subjective
- **B.** Objective
- **C.** Assessment
- **D.** Patient care

10. You will document the SAMPLE history in which section of the written patient report?
- **A.** Subjective
- **B.** Objective
- **C.** Assessment
- **D.** Patient care

Medication Administration

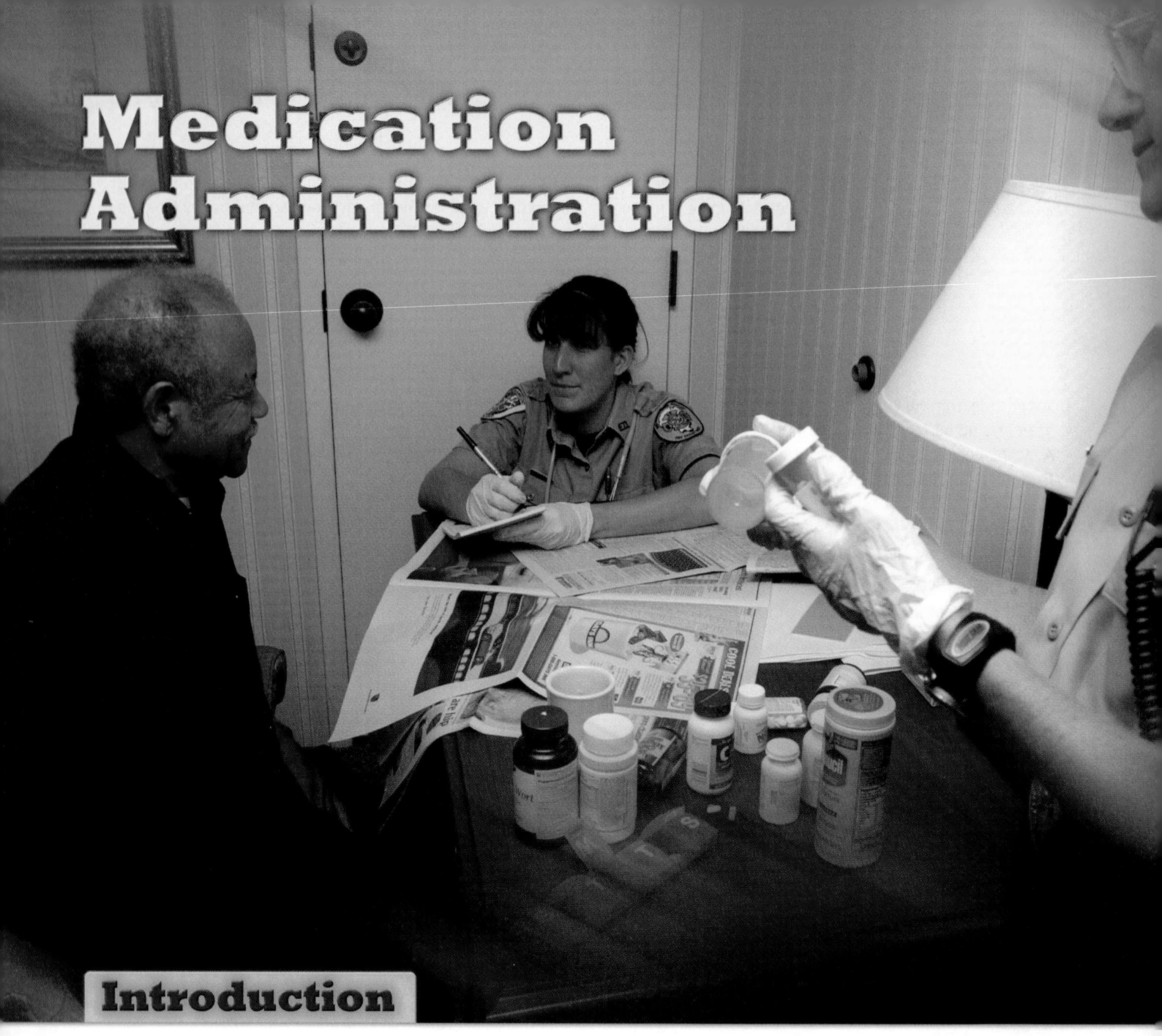

Introduction

Administration of medication to any patient carries with it additional responsibilities. Although medications can be lifesaving, they can also be dangerous. Even the simplest medication can have serious side effects if given to a person with an allergy to the drug. Other effects can occur with over dosage or under dosage.

It is essential for all responders to understand the indications, contraindications, complications, side effects, doses, and methods of delivery for every drug carried on the EMS unit. In many cases, EMS personnel are allowed to give medications only if the patient already has the drug or a prescription for the drug. Before giving any medication, by protocol or patient assistance, review the "six rights" of medication administration. This ensures the *Right drug*, is given to the *Right patient*, at the *Right time*, by the *Right route*, at the *Right dose*, and with the *Right documentation* in the patient care report. The skills in this section will prepare you to administer some common medications.

SECTION 4

Bronchodilator—Handheld Metered-Dose Inhaler

Performance Objective

Given an adult patient and appropriate equipment and supplies, the candidate shall recognize and select the medication ordered, properly assemble the delivery device, and administer the drug using the criteria herein prescribed, in 5 minutes or less.

Equipment

The following equipment is required to perform this skill:

- Appropriate body substance isolation/personal protective equipment
- Metered-dose inhaler (as prescribed to the patient or carried on the ambulance under approval of medical control)

Equipment that may be helpful:

- Spacer
- Oxygen cylinder and regulator
- Oxygen delivery device (nasal cannula, nonrebreathing mask)
- Stethoscope
- Pulse oximeter
- End-tidal carbon dioxide meter
- Sphygmomanometer (with various-sized cuffs)

Indications

- Respiratory distress from upper airway constriction
  - Asthma
  - Bronchiolitis
  - Chronic obstructive pulmonary disease (COPD)
  - Anaphylaxis

Contraindications

- Dependent upon medication used

Complications

- May be ineffective in patients with respiratory failure (maximum effort, minimal air movement)
- Not a "cure-all." Should be used appropriately

Procedures

Step 1 Ensure body substance isolation before beginning procedures.

Prior to beginning patient care, appropriate body substance isolation procedures should be employed.

Step 2 Avoid contamination of equipment or replace contaminated equipment prior to use.

Maintain aseptic technique throughout the skill performance. Keep visual contact with sterile equipment and ensure the sterile field is not compromised.

Step 3 Confirm order (medication, dosage, and route).

Repeat order given by physician. Repeat drug, dosage, route, and any additional special considerations. The first step is maintaining the *six rights* of drug administration: right drug, right patient, right dose, right route, right time, right documentation.

Step 4 Inform patient of order for medication and inquire about allergies and recent doses of other bronchodilators.

Inform the patient of the order received, the need for the therapy, and any concerns about its administration. Ask the patient if he or she has any allergies to the specific drug ordered, as well as to any other drugs, iodine, or pertinent foods.

Step 5 Select correct medication.

From the medication box, choose the correct medication as ordered by the physician.

Step 6 Check medication for contamination and expiration date.

Inspect the medication label for correct name and appropriate concentration, and the fluid for clarity, discoloration, and obvious contamination or signs of loss of sterility. Check that the expiration date has not passed.

Step 7 Shake the inhaler.

Actively shake the inhaler several times.

Step 8 Attach spacer to inhaler, if available.

If a spacer is available, attach it to the mouthpiece of the inhaler.

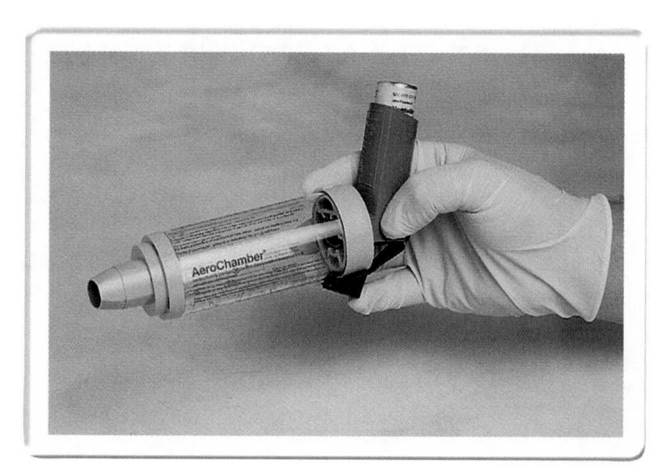

Step 9 Recheck the medication label.

Reinspect the medication label for correct name and appropriate concentration, and the fluid for clarity, discoloration, and obvious contamination or signs of loss of sterility. Check that the expiration date has not passed.

Step 10 Remove nonrebreathing mask from patient.

Remove the oxygen mask from the patient with the oxygen still flowing.

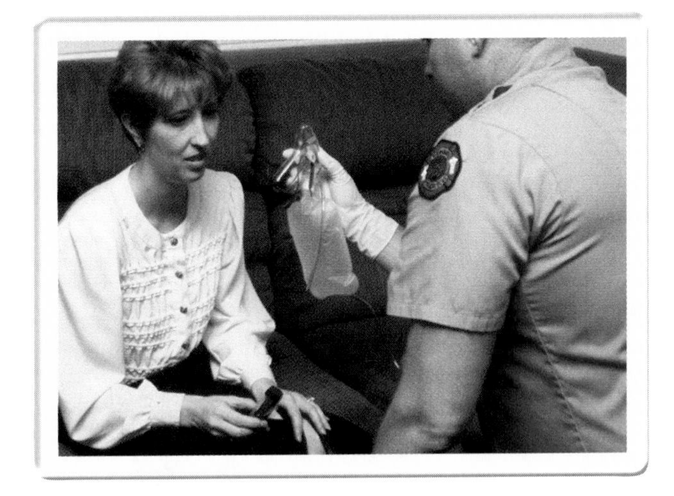

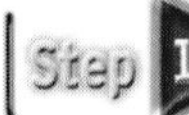

Step 11 Instruct patient to exhale deeply.

Explain to the patient how the inhaler works and have the patient take a deep breath and exhale completely.

Step 12 Instruct patient to put the mouthpiece in mouth and make a seal with lips.

Have the patient put the mouthpiece in his or her mouth and set the lips tightly.

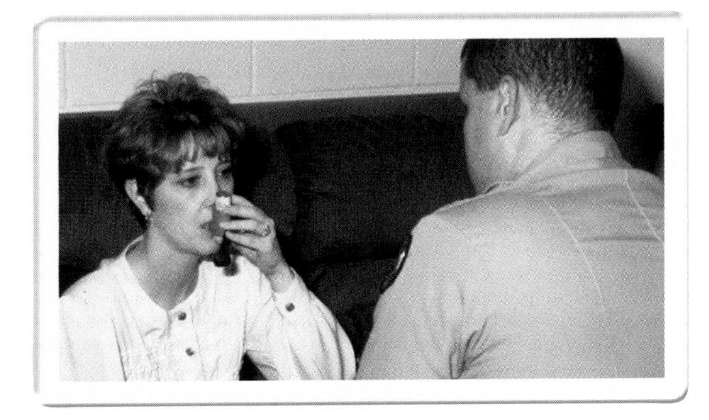

Step 13 Instruct patient to depress the inhaler canister while inhaling and then hold breath as long as comfortable.

Instruct the patient to inhale deeply as the canister is depressed. Encourage the patient to hold his or her breath as long as possible.

Step 14 Replace nonrebreathing mask on patient.

Reposition the nonrebreathing mask on the patient and ensure the liter flow is still at 15 L/min.

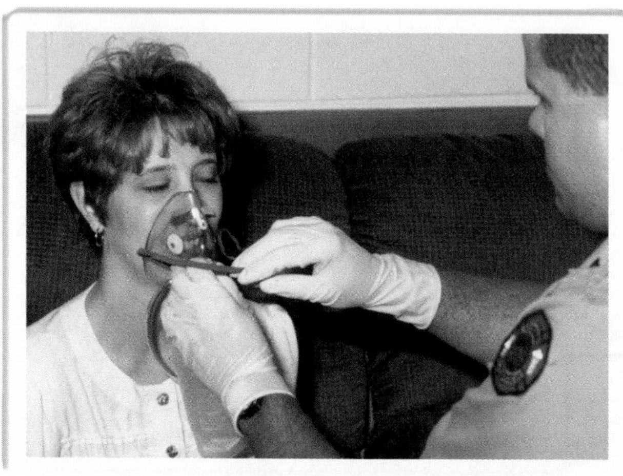

Step 15 Monitor for effects of medication.

Monitor the patient for the desired effects of medication administration by reassessing vital signs, tissue color, and difficulty of breathing. Document all changes, beneficial or not, that occur. Adjust management as needed.

Bronchodilator—Small-Volume Nebulizer

SKILL 24

Performance Objective

Given an adult patient and appropriate equipment and supplies, the candidate shall recognize and select the medication ordered, properly assemble the delivery device, and administer the drug using the criteria herein prescribed, in 5 minutes or less.

Equipment

The following equipment is required to perform this skill:

- Appropriate body substance isolation/personal protective equipment
- Oxygen cylinder and regulator
- Oxygen delivery device (nasal cannula, nonrebreathing mask)
- Small-volume nebulizer
- Bronchodilator (as prescribed to the patient or carried on the ambulance under approval of medical control)

Equipment that may be helpful:

- Stethoscope
- Pulse oximeter
- End-tidal carbon dioxide meter
- Sphygmomanometer (with various-sized cuffs)

Indications

- Respiratory distress from upper airway constriction
  - Asthma
  - Bronchiolitis
  - Chronic obstructive pulmonary disease (COPD)
  - Anaphylaxis

Contraindications

- Dependent upon medication used

Complications

- May be ineffective in patients with respiratory failure (maximum effort, minimal air movement)
- Not a "cure-all." Should be used appropriately

Procedures

This patient will have a nonrebreathing mask on at the beginning of the scenario.

Step 1 Ensure body substance isolation before beginning procedures.

Prior to beginning patient care, appropriate body substance isolation procedures should be employed.

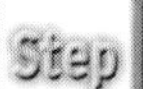 

Step 2 Avoid contamination of equipment or replace contaminated equipment prior to use.

Maintain aseptic technique throughout the skill performance. Keep visual contact with sterile equipment and ensure the sterile field is not compromised.

Step 3 Confirm order (medication, dosage, and route).

Repeat order given by physician. Be sure to repeat the drug, dosage, route, and any special considerations given. The first step is maintaining the *six rights* of drug administration: right drug, right patient, right dose, right route, right time, right documentation.

Step 4 Inform patient of order for medication and inquire about allergies and recent doses of other bronchodilators.

Inform the patient of the order received, the need for the therapy, and any concerns about its administration. Ask the patient if he or she has any allergies to the specific drug ordered, as well as to any other drugs, iodine, or pertinent foods.

Step 5 Select correct medication.

From the medication box, choose the correct medication as ordered by the physician.

Visually check medication for contamination and expiration date.

Visually inspect the medication label for correct name and appropriate concentration, and the fluid for clarity, discoloration, and obvious contamination or signs of loss of sterility. Check that the expiration date has not passed.

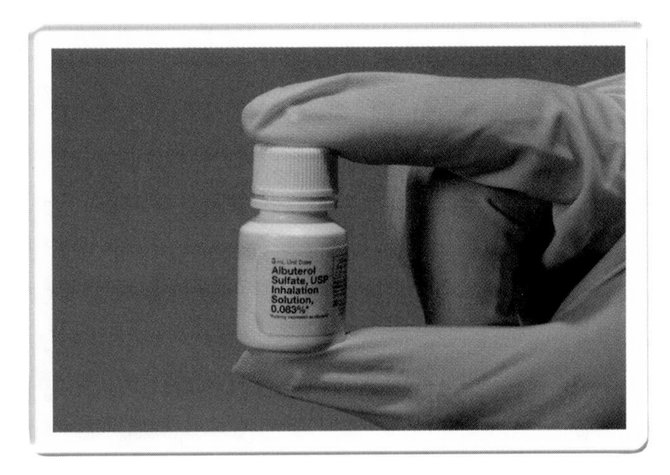

Add appropriate volume of medication to the nebulizer.

Open the nebulizer and add the ordered volume. This is normally the entire amount of the container, but mixing the medication with sterile saline or using partial doses may be necessary to achieve the optimal volume of fluid for nebulized function.

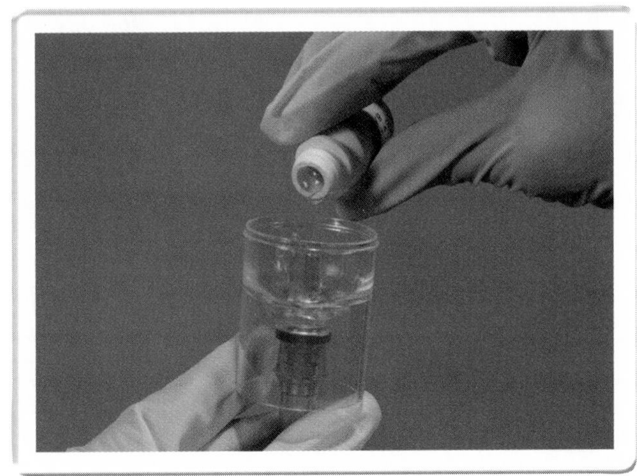

Step 8 Assemble nebulizer according to the manufacturer's standard (or local protocol) and connect to oxygen regulator.

Reassemble the nebulizer and attach the mouthpiece and exhaust tube. Connect the oxygen tubing to the nebulizer and the oxygen regulator.

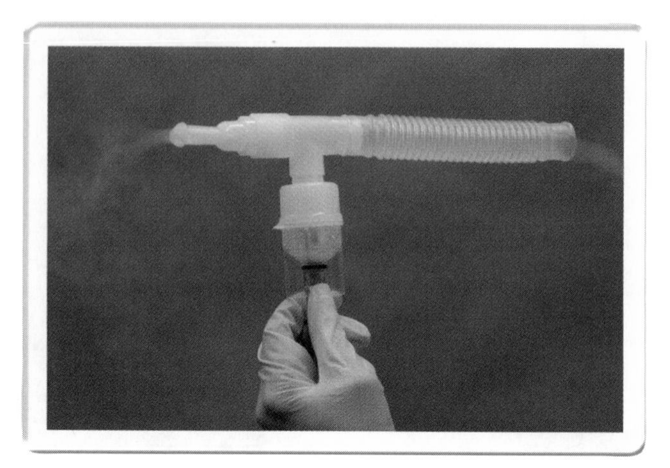

Step 9 Recheck the medication label.

Reinspect the medication label for correct name and appropriate concentration, and the fluid for clarity, discoloration, and obvious contamination or signs of loss of sterility. Check that the expiration date has not passed.

Step 10 Adjust oxygen flow as ordered and allow mist to fill breathing tube or mask prior to applying to patient.

Adjust the oxygen flow to achieve the desired misting effect (usually between 6 to 10 L/min) and wait for a slight mist to come from the end of the tube. If after several seconds the mist does not form, a slight increase in flow may be necessary.

Remove nonrebreathing mask and position nebulizer device on patient.

Remove the nonrebreathing mask and hand the nebulizer to the patient. Instruct the patient to place the nebulizer to his or her mouth and breathe normally, inhaling the mist. Reassess the patient, recording the respiratory effort, pulse, and breath sounds.

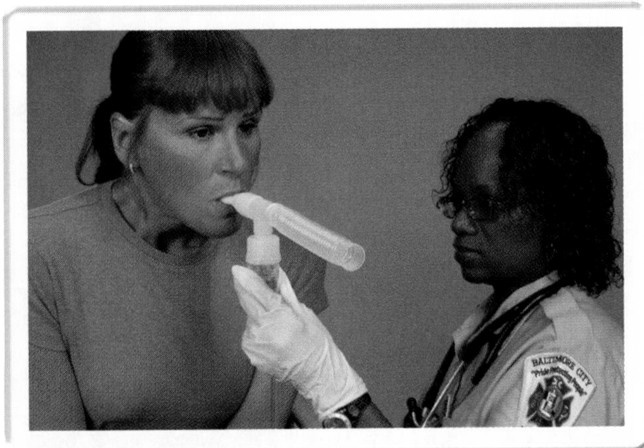

Monitor for effects of medication.

Monitor the patient for the desired effects of medication administration by reassessing vital signs, tissue color, and difficulty of breathing. Document all changes, beneficial or not, that occur. Adjust management as needed.

SKILL 25

Epinephrine Auto-Injectors

Performance Objective

Given an epinephrine auto-injector and appropriate equipment and supplies, the candidate shall be able to recognize and select the medication ordered, properly utilize the auto-injector, and administer the drug using the criteria herein prescribed, in 5 minutes or less.

Equipment

The following equipment is required to perform this skill:

- Appropriate body substance isolation/personal protective equipment
- Epinephrine auto-injector (as prescribed to the patient or carried on the ambulance under approval of medical control)

Equipment that may be helpful:

- Oxygen cylinder
- Oxygen regulator
- Oxygen delivery device (nasal cannula, nonrebreathing mask)
- Stethoscope
- Pulse oximeter
- End-tidal carbon dioxide meter
- Sphygmomanometer (with various-sized cuffs)

Indications

- Respiratory distress from severe allergic reactions

Contraindications

- Allergies to epinephrine

Complications

- May cause extreme tachycardia

Procedures

Ensure body substance isolation before beginning procedures.

Prior to beginning patient care, appropriate body substance isolation procedures should be employed.

Avoid contamination of equipment or replace contaminated equipment prior to use.

Maintain aseptic technique throughout the skill performance. Keep visual contact with sterile equipment and ensure the sterile field is not compromised.

Confirm order (medication, dosage, and route).

Repeat order given by physician: drug, dosage, route, and any special considerations given. The first step is maintaining the *six rights* of drug administration: right drug, right patient, right dose, right route, right time, right documentation.

Inform patient of order for medication and inquire about allergies.

Inform the patient of the order received, the need for the therapy, and any concerns about its administration. Ask the patient if he or she has any allergies to the specific drug ordered, as well as to any other drugs, iodine, or pertinent foods.

CHECK FOR IODINE ALLERGIES

Select correct medication.

From the medication box, choose the correct medication as ordered by the physician.

In the Field

Use of Epinephrine Auto-Injectors

EpiPens deliver 0.3 mg of epinephrine and are appropriate for all adults and for children weighing 25 kg or more. The EpiPen, Jr delivers 0.15 mg and is appropriate for children weighing 15 to 25 kg.

Step 6 Visually check medication for contamination and expiration date.

Visually inspect the medication label for correct name and appropriate concentration, and the fluid for clarity, discoloration, and obvious contamination or signs of loss of sterility. Check that the expiration date has not passed.

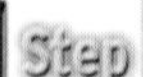 

Step 7 Select appropriate site.

Locate the correct site for injection, usually the lateral midthigh.

Step 8 Recheck the medication label.

Reinspect the medication label for correct name and appropriate concentration, and the fluid for clarity, discoloration, and obvious contamination or signs of loss of sterility. Check that the expiration date has not passed.

Step 9 Prepare the injection site.

Choose an injection site in the large muscle of the thigh. If time permits, cleanse the injection site to kill microbial agents that live on the patient's skin. In severe cases of anaphylaxis, it is acceptable to inject the epinephrine through the patient's pants.

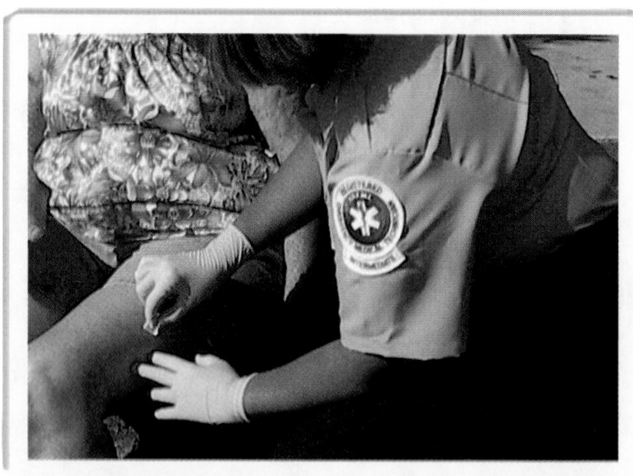

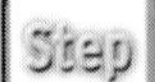 

Step 10 Remove safety cap from the auto-injector.

Remove the gray end cap (on the thick end) from the auto-injector.

Step 11 Place the tip of the auto-injector against the injection site and push the injector firmly against the injection site.

Place the black tip (narrow end) of the auto-injector against the intended injection site. Push firmly and quickly until the needle is released and injects into the patient.

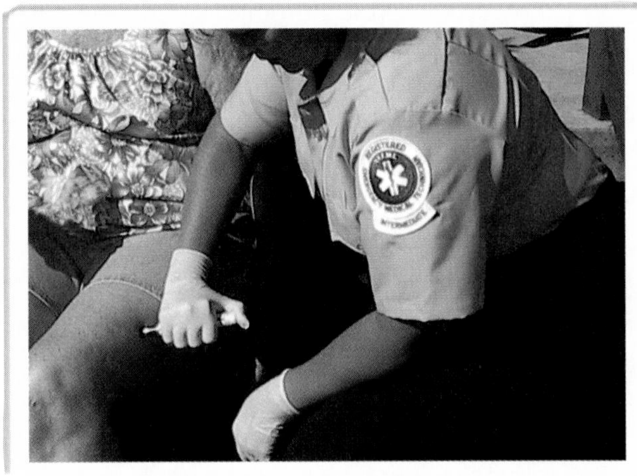

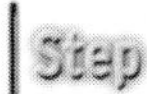 

Step 12 Hold auto-injector against the site for 10 seconds.

Hold the auto-injector against the injection site for at least 10 seconds.

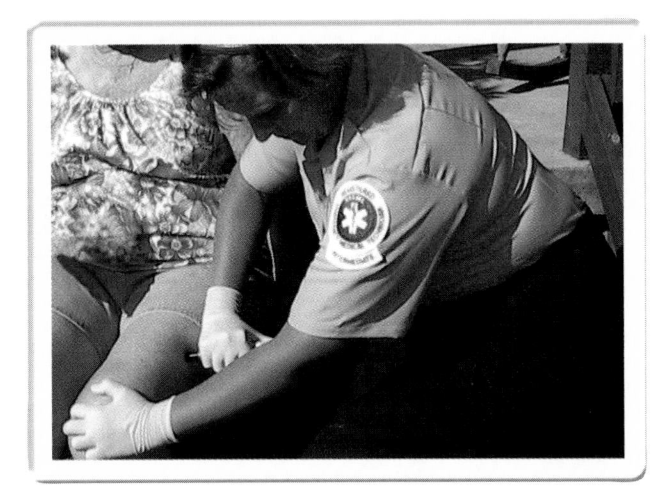

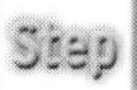

Step 13 Remove auto-injector and apply pressure.

Quickly remove the auto-injector by pulling it straight out, and apply pressure with a 4″ x 4″ gauze dressing to the injection site. Remember that the auto-injector will have a contaminated needle sticking from the black end.

Step 14 Dispose of contaminated equipment.

Dispose of the auto-injector by inserting the entire device into a puncture-resistant sharps container, needle first. Use extreme caution with the exposed dirty needle of the auto-injector until it is disposed of.

Step 15 Monitor for effects of medication.

Monitor the patient for the desired effects of medication administration by reassessing vital signs, tissue color, and difficulty of breathing. Document all changes, beneficial or not, that occur. Adjust management as needed.

Oral Medication Administration

SKILL 26

Performance Objective

Given all necessary medications and equipment, the candidate shall be able to recognize and select the medication ordered and administer the drug using the criteria herein prescribed, in 2 minutes or less.

Equipment

The following equipment is required to perform this skill:

- Appropriate body substance isolation/personal protective equipment
- Appropriate medication (as prescribed to the patient or carried on the ambulance under approval of medical control)

Equipment that may be helpful:

- Tongue depressor
- Disposable cup
- Drinking straw
- Stethoscope
- Pulse oximeter
- End-tidal carbon dioxide meter
- Sphygmomanometer (with various-sized cuffs)

Indications

- Specific to the medication given

Contraindications

- Patients with altered level of consciousness or who are unresponsive

Complications

- Medication specific

Procedures

Oral medication administration may include giving aspirin, nitroglycerin, or glucose paste.

Step 1 Ensure body substance isolation before beginning procedures.

Prior to beginning patient care, appropriate body substance isolation procedures should be employed.

Safety Tips

Inquiring About Drugs and Allergies

It is essential that a complete history be performed before giving a patient any medication. An important question that must be asked before administering nitroglycerin is whether the patient is using medications for erectile dysfunction. Administration of nitroglycerin within 48 hours of a patient taking sildenafil (Viagra) and similar drugs can create serious complications, including death.

Step 2 Avoid contamination of equipment or replace contaminated equipment prior to use.

Maintain aseptic technique throughout the skill performance. Keep visual contact with sterile equipment and ensure the sterile field is not compromised.

Step 3 Confirm order (medication, dosage, and route).

Repeat order given by physician: drug, dosage, route, and any special considerations given. The first step is maintaining the *six rights* of drug administration: right drug, right patient, right dose, right route, right time, right documentation.

Step 4 Inform patient of order for medication and inquire about allergies.

Inform the patient of the order received, the need for the therapy, and any concerns about its administration. Ask the patient if he or she has any allergies to the specific drug ordered, as well as to any other drugs, iodine, or pertinent foods.

Safety Tips

Understanding Medications

Each drug carried on your EMS unit has specific uses. It is *essential* for all EMS personnel who carry drugs on their unit to understand all the indications, contraindications, side effects, precautions, doses, and routes of administration for each drug carried.

Step 5 Select correct medication.

From the medication box, choose the correct medication as ordered by the physician.

Step 6 Visually check medication for contamination and expiration date.

Visually inspect the medication label for correct name and appropriate concentration, and the fluid for clarity, discoloration, and obvious contamination or signs of loss of sterility. Check that the expiration date has not passed.

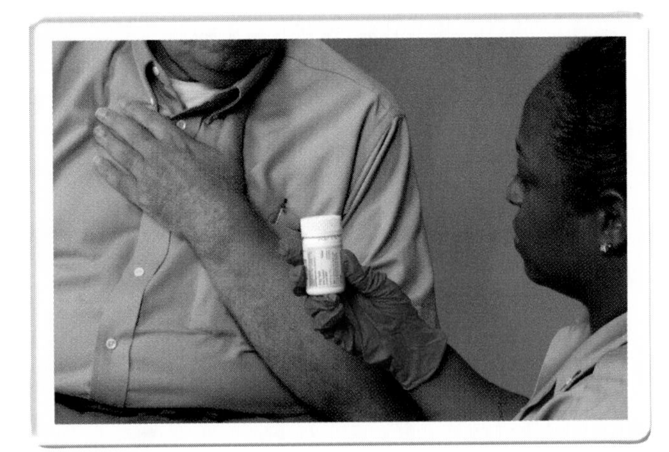

Safety Tips

Sublingual Nitroglycerin Spray

Nitroglycerin is a stable compound that will not explode under normal circumstances. However, there are some usage and safety rules that need to be observed. Do not shake the drug container before administration. Ensure that the opening of the spray nozzle is aimed into the patient's mouth, onto or under the tongue. The patient should not swallow or inhale the spray. It is advisable for EMS personnel to wear gloves while administering to avoid nitroglycerin getting on the skin. Even small amounts of nitroglycerin will be absorbed and can result in severe headaches and drops in blood pressure.

Step 7 Ensure the patient is conscious and can maintain his or her own airway.

Before administering oral medications, ensure the patient has the ability to swallow and maintain his or her own airway.

Step 8 Place, or instruct patient to place, medication into oral cavity appropriately.

Place the medication in the oral cavity using the following criteria:

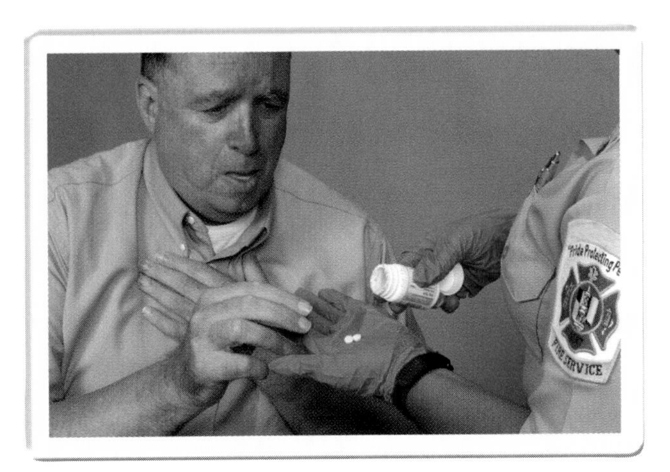

continued

- **Aspirin** should be swallowed. Some EMS systems allow the use of a small amount of water to assist the patient in swallowing the tablet; others require the tablet to either be swallowed dry or to be chewed.
- **Activated charcoal** is administered as a slurry that the patient will drink. Some brands of activated charcoal come as a prepared slurry, whereas others may require you to mix dry activated charcoal with water, magnesium citrate, or another liquid as described in protocols. Place the appropriate amount of the activated charcoal slurry into a cup or other acceptable container. Have the patient drink the activated charcoal either directly or through a straw.
- **Glucose paste** should be given by placing the paste on the end of a tongue depressor and then placing the depressor in the mouth between the cheek and gum. Allow the paste to be absorbed.
- **Nitroglycerin** should be placed under the patient's tongue. Ask the patient to open his or her mouth and lift the tongue. Place the nitroglycerin tablet under the tongue. Have the patient keep the tablet in place until it is dissolved.

Monitor for effects of medication.

Monitor the patient for the desired effects of medication administration by reassessing vital signs, tissue color, change in pain, or change in level of consciousness. Document all changes, beneficial or not, that occur. Adjust management as needed.

Check Your Knowledge

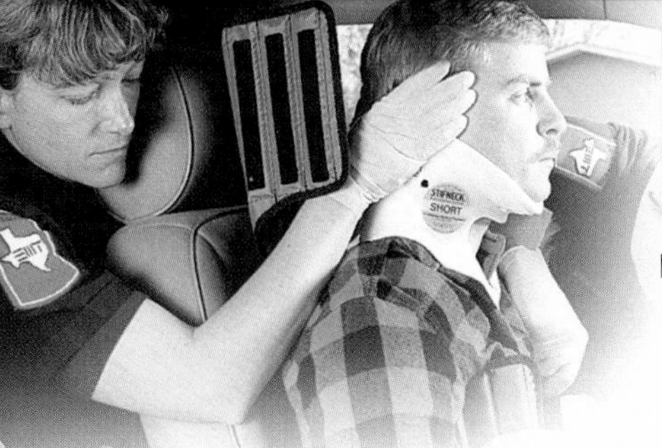

You are dispatched to a retail business on Main Street for difficulty breathing. You arrive to find a 46-year-old woman sitting on a crate in the back of the store in obvious respiratory distress. She tells you that she was cleaning out the storage room when she suddenly became short of breath.

1. **Which of the following conditions indicates the use of a handheld metered-dose inhaler?**
   - **A.** Hypertension
   - **B.** Asthma
   - **C.** Pneumonia
   - **D.** Congestive heart failure

2. **Under what circumstances may you use the patient's coworker's metered-dose inhaler for the patient?**
   - **A.** The patient's next of kin gives you permission to do so.
   - **B.** The medical director gives you on-line direction to do so.
   - **C.** The patient tells you it is the same prescription he or she regularly uses.
   - **D.** You may only administer a metered-dose inhaler to the specific individual for which it has been prescribed.

3. **You should administer the metered-dose inhaler in ____ minutes or less.**
   - **A.** 2
   - **B.** 3
   - **C.** 5
   - **D.** 10

4. **Before administering any medication, you should confirm that you have the right drug, right patient, right dose, and right _________.**
   - **A.** route
   - **B.** day
   - **C.** doctor
   - **D.** all of the above

5. **After administration of a metered-dose inhaler, monitor the effects of the medication including:**
   - **A.** vital signs.
   - **B.** tissue coloring.
   - **C.** difficulty breathing.
   - **D.** all of the above.

6. **Before using a small-volume nebulizer, it is important to ask the patient:**
   - **A.** if he or she has taken any recent doses of other bronchodilators.
   - **B.** if he or she has any allergies.
   - **C.** if he or she has any concerns about administration.
   - **D.** all of the above.

7. **To achieve the optimal volume of fluid for nebulizer function, it may be necessary to:**
   - **A.** mix the medication with sterile saline.
   - **B.** give additional doses.
   - **C.** try an alternate medication.
   - **D.** all of the above.

8. **What is the primary potential complication from administration of an epinephrine auto-injector?**
   - **A.** Difficulty breathing
   - **B.** Low blood pressure
   - **C.** Extreme tachycardia
   - **D.** Diaphoresis

9. **When administering an epinephrine auto-injector, you must hold the auto-injector against the injection site for ______ seconds.**
   - **A.** 5
   - **B.** 10
   - **C.** 15
   - **D.** 20

10. **Administration of oral medication is contraindicated in patients who are:**
    - **A.** under the age of 12.
    - **B.** having difficulty breathing.
    - **C.** hypertensive.
    - **D.** unresponsive or with an altered level of consciousness.

Trauma Management

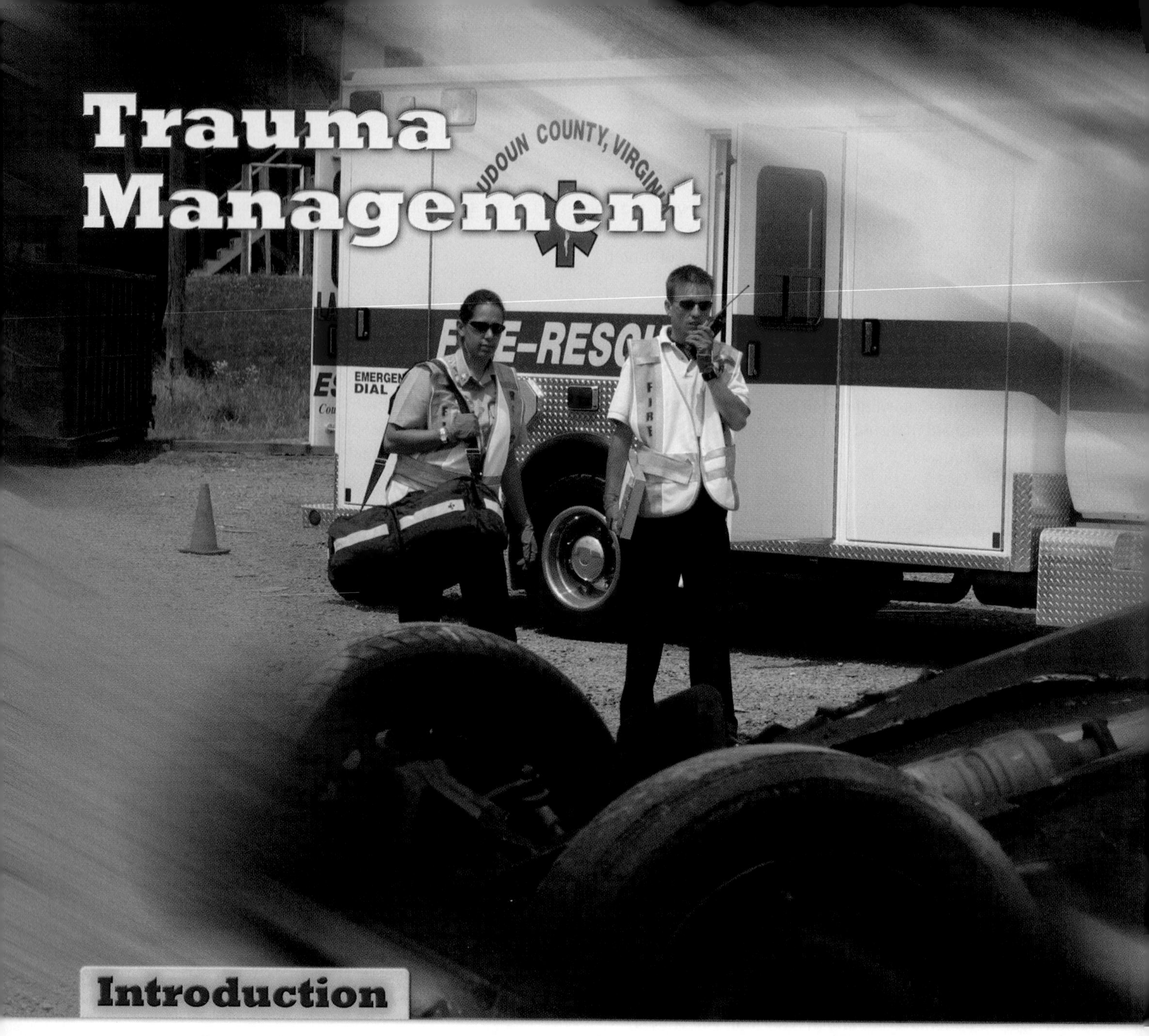

Introduction

Trauma is the number one killer of children and young adults. Trauma management skills are important in the control of bleeding and in the stabilization of spinal injuries, fractures, and dislocations.

The most important trauma skill in the management of a trauma patient is often rapid transport. Although the management of bleeding and the stabilization of spinal injuries and fractures are a part of total patient care, delaying transport in order to perform such skills can be a fatal error in patient care. Always weigh the importance of patient care against the immediate needs of the patient. In general, trauma care for a patient with serious injuries should involve assurance of an airway, management of serious hemorrhage, stabilization of the cervical and thoracic spine, and rapid transport. Bleeding control and splinting can be deferred or performed en route to the hospital.

It is important to learn how to perform each of the skills in this section, and to learn when they should be used. Proper management of trauma patients is essential in reducing death and disability.

SECTION 5

SKILL 27

General Bandaging Techniques

Performance Objective

Given a patient and a description of the patient's injuries, the candidate shall apply a dressing and bandage according to the principles of hemorrhage control and aseptic technique, in 10 minutes or less.

Equipment

The following equipment is required to perform this skill:

- Appropriate body substance isolation/personal protective equipment
- 4" x 4" gauze dressings
- 5" x 9" dressings
- 3" roller bandages
- 6" roller bandages
- Triangular bandages
- 1" tape (cloth or silk)

Equipment that may be helpful:

- Multitrauma dressings
- Elastic roller bandages
- Sterile saline (500 mL and 1,000 mL bottles)

Indications

- Control of external hemorrhage

Contraindications

- None

Complications

- Tight bandages may lead to constriction of distal circulation.

Procedures

Ensure body substance isolation before beginning procedures.

Prior to beginning patient care, appropriate body substance isolation procedures should be employed.

Check circulation (capillary refill or pulse) distal to injury before bandaging.

A check of the distal circulation through capillary refill or pulse check must be performed when appropriate and possible.

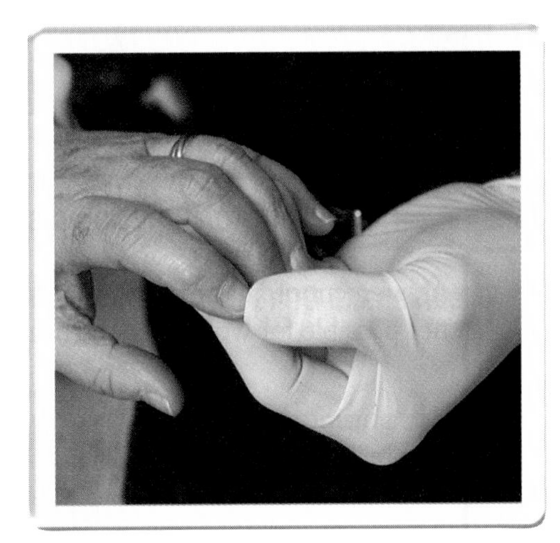

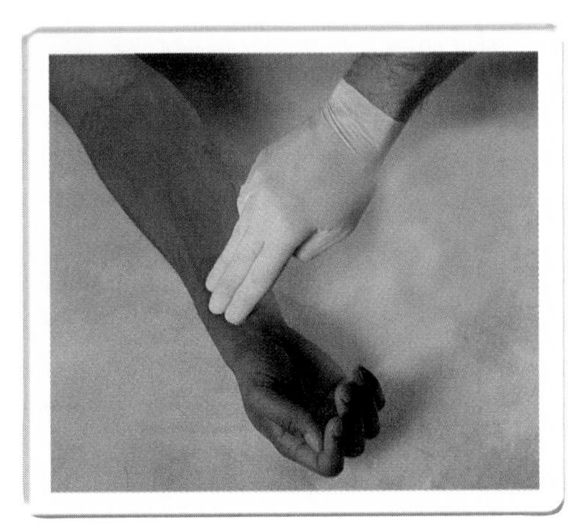

Step 7 Use bandaging technique appropriate to injury.

Technique must control hemorrhage without aggravation of the injury.

See suggested procedures in Skill 29 for techniques appropriate for various types of injuries.

Step 8 Reassess circulation (capillary refill or pulse), sensation, and motor function distal to injury after bandaging.

A check of the distal circulation through capillary refill or pulse check must be performed when appropriate and possible. Reassessment of motor function and sensation must be performed on all extremities. This step is not possible for wounds to the head, eye, or neck or in situations of amputation.

Bleeding Control and Shock Management

Performance Objective

Given a patient and a description of the patient's injuries, the candidate shall apply the principles of hemorrhage control and shock management, in 10 minutes or less.

Equipment

The following equipment is required to perform this skill:

- Appropriate body substance isolation/personal protection equipment
- 4″ x 4″ gauze dressings
- 5″ x 9″ dressings
- 3″ roller bandages
- 6″ roller bandages
- Triangular bandages
- 1″ tape (cloth or silk)
- Blanket

Equipment that may be helpful:

- Multitrauma dressings
- Elastic roller bandages
- Sterile saline (500 mL and 1,000 mL bottles)

Indications

- Control of external hemorrhage and shock

Contraindications

- None

Complications

- None

Procedures

This is a sequential procedure. The actual care of the patient will dictate which of the following steps are included. Assessment of circulation, sensation, and motor function distal to the injury should be performed at some point in this sequence (as described in Skill 27). The actual point at which this assessment takes place will depend on the severity of the situation. Less severe bleeding will allow more opportunities for distal assessment than life-threatening hemorrhage.

Step 1 Ensure body substance isolation before beginning procedures.

Prior to beginning patient care, appropriate body substance isolation procedures should be employed.

Step 2 Identify wound and apply dressing and direct pressure.

Upon identification of a bleeding wound, immediately apply a dressing and direct pressure.

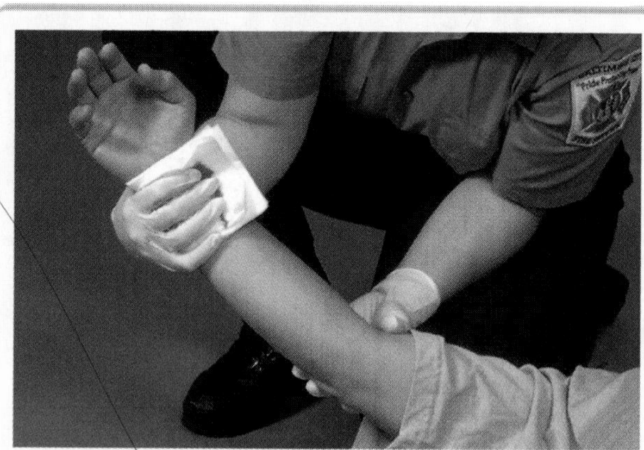

Step 3 Elevate extremity.

Elevate the extremity while maintaining direct pressure.

Note: If the wound continues to bleed, continue to Step 4. If bleeding is controlled, proceed to Step 6.

Apply an additional dressing to the wound without removing the first dressing.

If the wound continues to bleed, a bulk dressing may be needed. Leaving the first dressing in place, apply additional dressings to the wound. Continue to apply direct pressure with elevation.

Note: If the wound continues to bleed, continue to Step 5. If bleeding is controlled, proceed to Step 6.

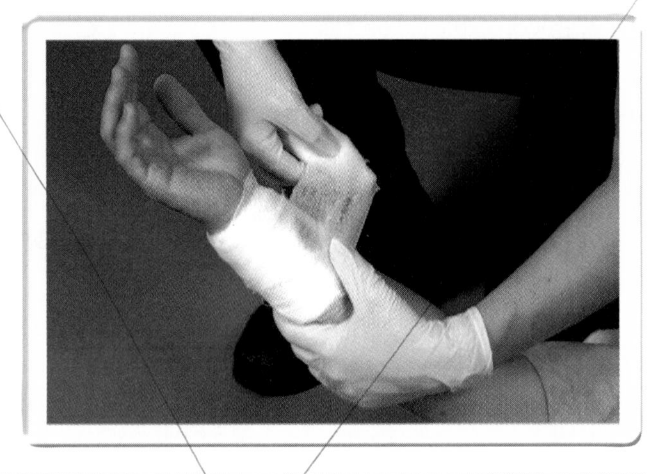

Step 5 Locate and apply pressure to appropriate arterial pressure point.

If bleeding is still not controlled, locate the artery above the wound. Apply pressure to the pressure point. If the initial pressure point fails to control bleeding, a more proximal artery should be tried.

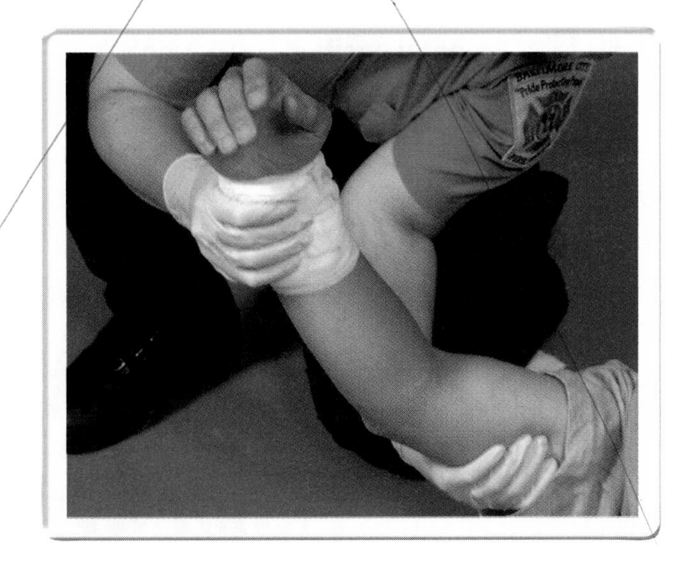

Step 5 continued

As a last resort to stop bleeding, a tourniquet may be applied. If a tourniquet is deemed necessary, it should be applied quickly and *not* released until after a physician has repaired the wound. Many rescuers use a blood pressure cuff as a tourniquet. This is acceptable as long as the cuff pressure can be maintained. A slow leak in cuff pressure can be severely damaging to the patient.

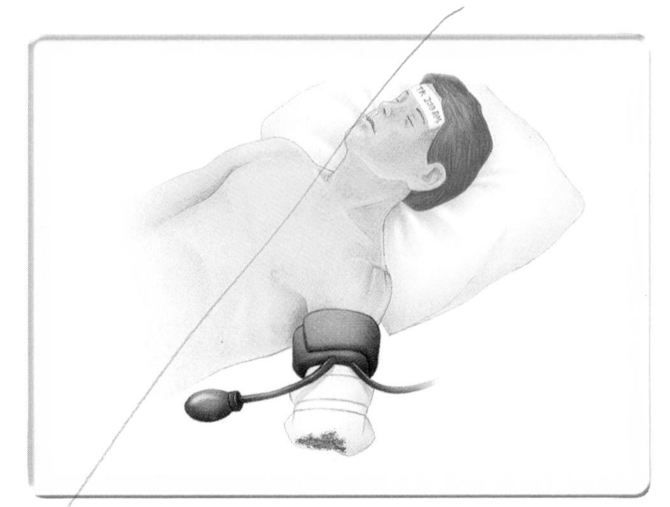

Step 6 Bandage the wound.

Once bleeding is controlled, cover the dressing completely with an appropriate bandaging technique.

Note: Perform a patient assessment to evaluate the patient's condition. If the patient presents with signs and symptoms of hypovolemia, proceed to Step 7.

Step 7 Properly position the patient.

Place the patient supine and elevate the legs. In conscious patients, or patients who are breathing on their own, a modified Trendelenburg position is preferred over a true Trendelenburg position.

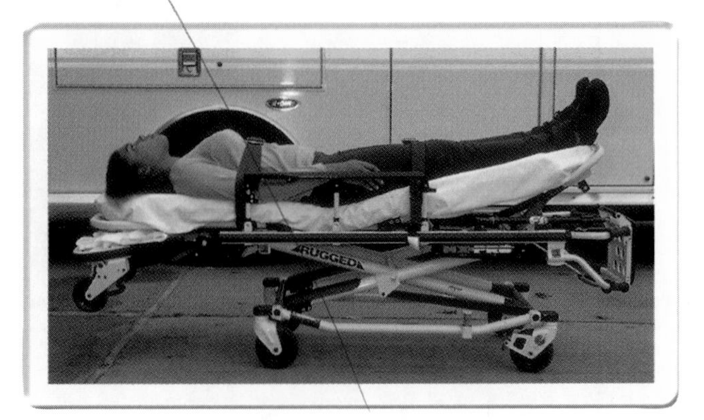

Step 8 Apply high-concentration oxygen.

Place the patient on high-flow, high-concentration oxygen via a nonrebreathing mask.

Step 9 Initiate steps to prevent heat loss from the patient.

Place a blanket over the patient to keep the patient from losing body heat. Do not assume that outside temperatures are high enough that a blanket is not needed.

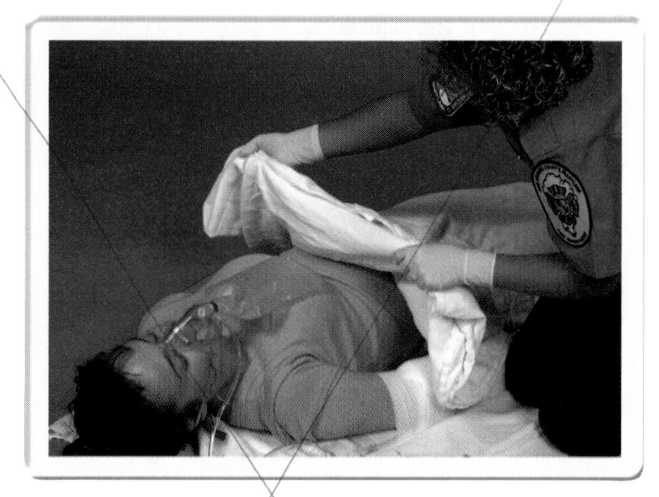

Step 10 Begin immediate transportation.

Hypovolemic shock is a serious medical condition. Immediate transportation to a trauma center is essential.

REFER TO SKILLS MANUAL

SKILL 29

Bandaging Techniques for Specific Injuries

Performance Objective

Candidates should be able to perform the following bandaging techniques. Alterations to these techniques are permitted provided the objectives of the skill are met.

Equipment

The following equipment is required to perform this skill:

- Appropriate body substance isolation/personal protective equipment
- 4″ x 4″ gauze dressings
- 5″ x 9″ dressings
- 3″ roller bandages
- 6″ roller bandages
- Triangular bandages
- 1″ tape (cloth or silk)

Equipment that may be helpful:

- Multitrauma dressings
- Elastic roller bandages
- Sterile saline (500 mL and 1,000 mL bottles)

Indications

- None

Contraindications

- None

Complications

- None

Procedures

Step 1 Ensure body substance isolation before beginning procedures.

Prior to beginning patient care, appropriate body substance isolation procedures should be employed.

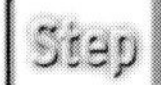

Step 2 Evaluate distal circulation (capillary refill or pulse), sensation, and motor function distal to injury as required (see Skill 27).

As appropriate and depending on the location of the injury, assess circulation, sensation, and motor function distal to the injury site.

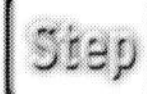

Step 3 Apply appropriate dressings and bandages specific to the injury.

Using aseptic technique, apply sterile dressings and bandages to the injury. Procedures specific to common injuries are detailed in the following subsections.

Scalp Laceration (Without Skull Fracture)

Because of the injury location, no evaluation of distal circulation, sensation, or motor function is required.

Using aseptic technique, apply a sterile dressing to the wound. Apply several more pieces of dressing material on top of the dressing to add bulk. Apply direct pressure. The direct pressure must be continued until the bandage is applied and secured.

Make a 1-inch fold along the base (long side) of a triangular bandage, and fold to the outside. Place the bandage over the top of the patient's head with the base positioned level with the eyebrows and the apex (point) behind the head.

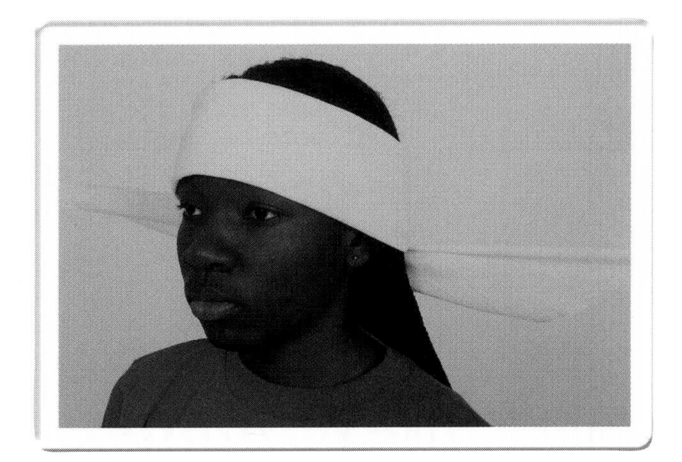

continued

Pull the two ends tightly around the back of the head, making sure the apex is underneath, and tie a half knot low on the back of the head.

If the tails are long enough, return the two ends to the front of the head and tie a square knot. Tuck the ends to keep them out of the patient's eyes.

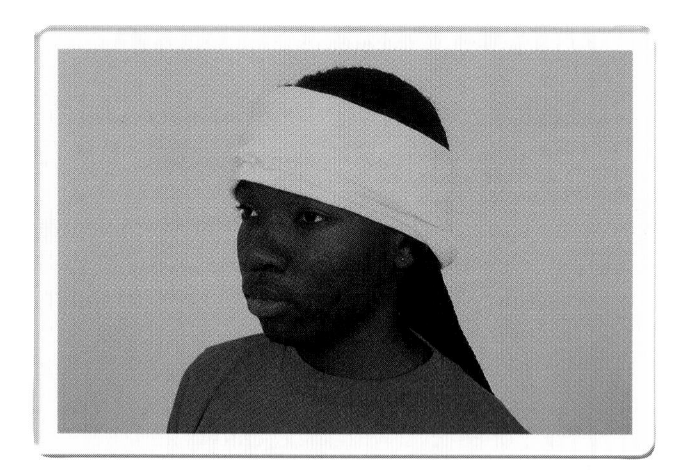

continued

Scalp Laceration (With Skull Fracture)

Because of the injury location, no evaluation of distal circulation, sensation, or motor function is required.

Using aseptic technique, apply a sterile dressing to the wound. Apply several more pieces of dressing material on top of the dressing to add bulk without placing direct pressure on the site. The added bulk must be sufficient to control hemorrhage without using direct pressure. It may be necessary to place large pieces of roller gauze bandage on the first dressing.

While holding the dressing materials in place, wrap a few turns of roller bandage around the head from forehead to back and anchor in place.

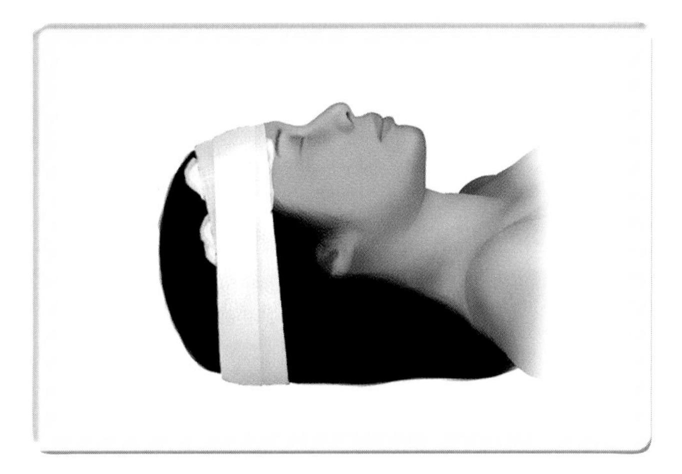

With recurrent turns, completely cover the top of the head and all dressing materials, including bulk. Secure in place using roller bandage applied around the forehead and back of head.

Eye Injury

Because of the injury location, no evaluation of distal circulation, sensation, or motor function is required.

All injuries that involve the globe of the eye should be covered with a moist, sterile dressing and bandaged without using pressure. Cups, pediatric oxygen masks, or commercial eye shields should be used to avoid pressure of any kind.

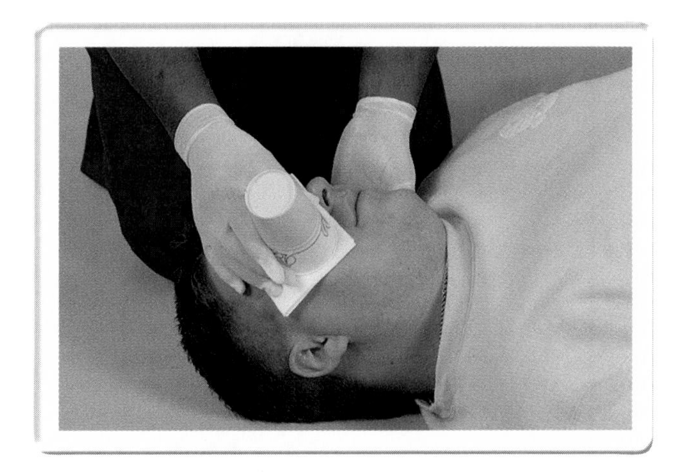

continued

Both the injured and uninjured eye should be covered to avoid sympathetic movement of the injured eye as the uninjured eye tracks. Be sure not to bandage over the patient's nose or mouth.

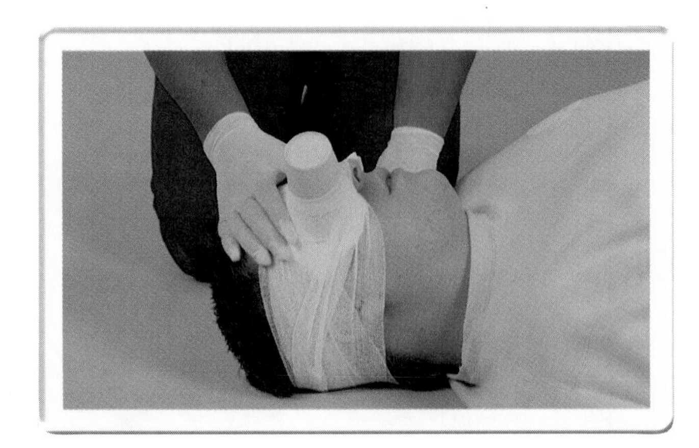

Facial Cheek Laceration

Because of the injury location, no evaluation of distal circulation, sensation, or motor function is required.

Using aseptic technique, apply a sterile dressing over the cheek laceration. Have the patient open his or her mouth to check for a through-and-through laceration. If the laceration continues into the oral cavity, place a sterile dressing into the mouth, between the cheek and gum, leaving a corner of the dressing visible. This is important to prevent aspiration by the patient. Additional dressing materials may be applied to the external dressing to form a bulk dressing.

Holding the dressing in place, wrap a few turns of roller bandage around the forehead and back of head and anchor. On successive turns, bring the roller bandage over the dressing materials and pull tightly to achieve a pressure dressing.

Avoid covering the patient's mouth or nose. When bandages are properly applied, the patient should not have any bandaging materials below the eyebrow of the uninjured side. All dressing materials should be covered by the roller bandage.

continued

Neck Laceration

Distal circulation should be assessed by checking the patient's level of consciousness and the carotid pulse above the injury site on the ipsilateral side.

Using aseptic technique, apply a sterile occlusive dressing (to prevent air embolism) to the laceration.

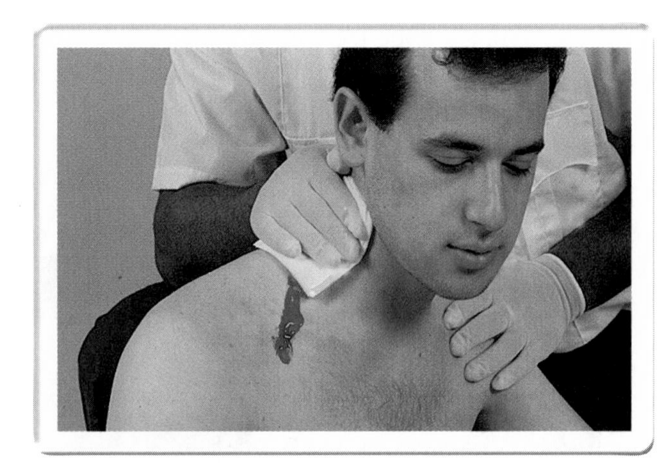

Secure with roller bandage wrapped around the injured side of the neck and the uninjured (opposite) armpit.

Reassess circulation and check that the airway has not been compromised.

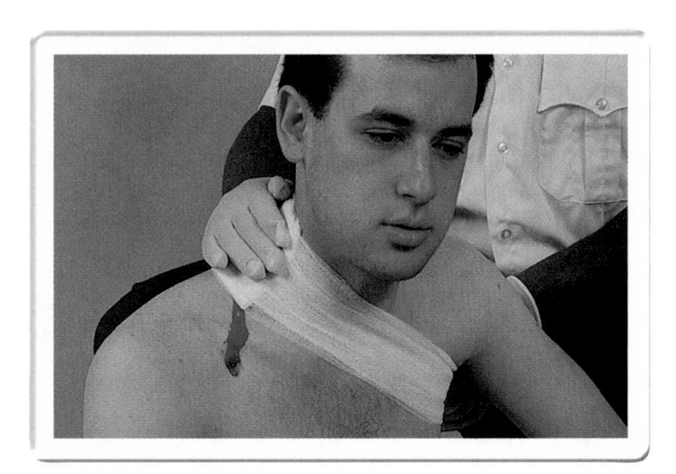

Step 3 continued

Sucking Chest Wound

Distal circulation to sucking chest wounds should be assessed any time the wound is superior to the nipple line on the anterior chest. Circulation, sensation, and motor function should be assessed in the arm of the injured side; however, due to the seriousness of this condition, this assessment should be delayed until after the sucking chest wound has been treated.

Using aseptic technique, apply a sterile occlusive dressing to the laceration. Place a gauze dressing over the occlusive dressing and bandage in place on the superior, medial, and inferior sides. Leave the lateral side open to act as a flutter valve, allowing excess chest pressure to escape.

Watch your patient carefully. Should severe respiratory distress or other signs of a tension pneumothorax develop, remove the dressing and bandage to release the thoracic pressure. Once the tension pneumothorax has been reduced, replace the dressing and the three-sided bandage.

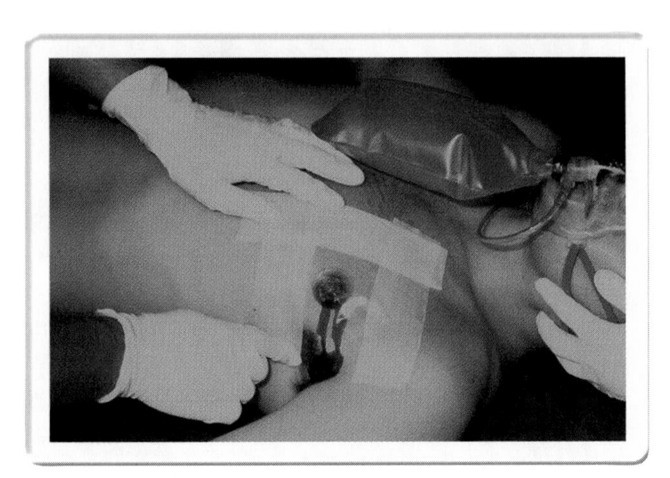

Joint Laceration

Assess circulation, sensation, and motor function distal to the injury site.

Using aseptic technique, apply a sterile dressing over the wound, covering the entire laceration. Place the extremity in a position of function (for instance, elbows should be flexed, placing the hand above the level of the elbow) and secure the dressing using a figure-of-8 (criss-cross) bandage with roller gauze.

Always wrap extremities from distal to proximal, using enough pressure to control bleeding without compromising circulation. All dressing material should be covered by the roller gauze.

Lacerations to upper extremities should be placed in a sling and secured with a swathe to avoid aggravation of the injury and for patient comfort (see Skill 32 for immobilization techniques). Reassess circulation, sensation, and motor function upon completion.

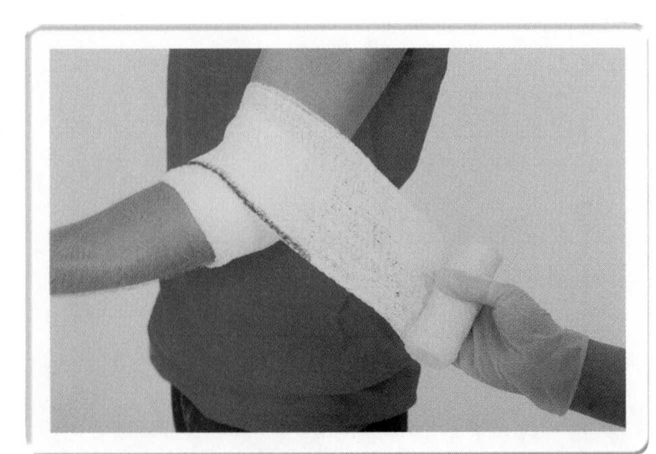

continued

Limb Laceration

Assess circulation, sensation, and motor function distal to the injury site.

Using aseptic technique, apply a sterile dressing over the wound, covering the entire laceration. Place the extremity in a position of function (for instance, elbows should be flexed, placing the hand above the level of the elbow) and secure the dressing using roller gauze.

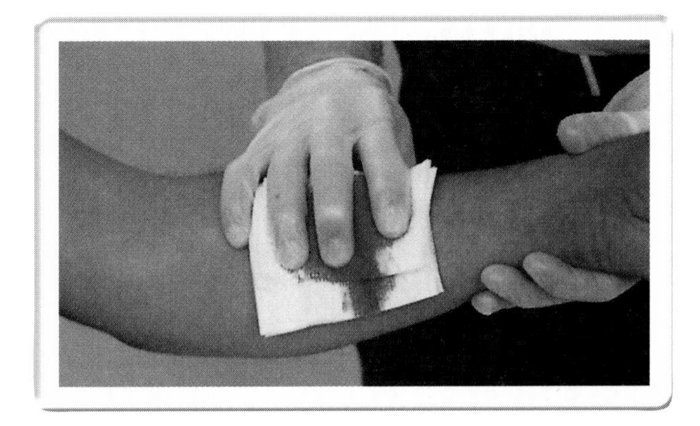

Always wrap extremities from distal to proximal, using enough pressure to control bleeding without compromising circulation. All dressing material should be covered by the roller gauze.

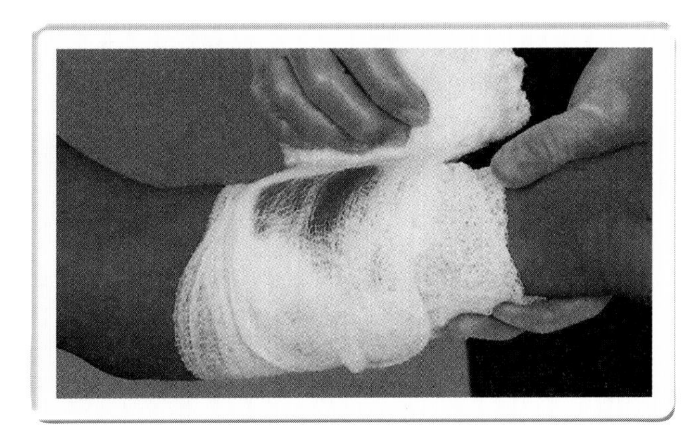

Lacerations to upper extremities should be placed in a sling and secured with a swathe to avoid aggravation of the injury and for patient comfort. Reassess circulation, sensation, and motor function upon completion.

Step 3 continued

Open (Compound) Fracture

The dressing and bandaging of the laceration associated with a compound fracture always takes precedence over the immobilization of the fracture.

Assess circulation, sensation, and motor function distal to the injury site.

Using aseptic technique, apply a dry sterile dressing over the wound, covering the entire laceration and all exposed bone. Place additional dressing materials if necessary to form bulk. If bone ends are protruding from the wound, place roller bandage on the medial and lateral side of the wound, running parallel to the bone(s). Secure the dressing, bulk, and roller bandage in place using triangular bandages applied in the following manner.

Place the body of the triangular bandage over the dressing and bulk with the apex of the bandage distal to the injury. Carefully tuck the tails of the triangular bandage under the injured extremity. Pull the tails of the triangular bandage snug and tie proximal to the injury using a square knot. Wrap a second triangular bandage around the distal end of the first bandage to secure the apex of the bandage.

Reassess circulation, sensation, and motor function before proceeding with the immobilization of the fracture (see Skill 32 for immobilization techniques).

Impaled Object

Impaled objects should never be removed from the body unless they are to the facial cheek or the object interferes with the performance of cardiopulmonary resuscitation (CPR). Impaled objects in the facial cheek may be removed after inspection inside the mouth reveals that the penetrating end of the object has not penetrated other structures, and that it is a potential cause of airway obstruction. If an object must be removed to perform adequate CPR, pack the wound with gauze to tamponade the bleeding.

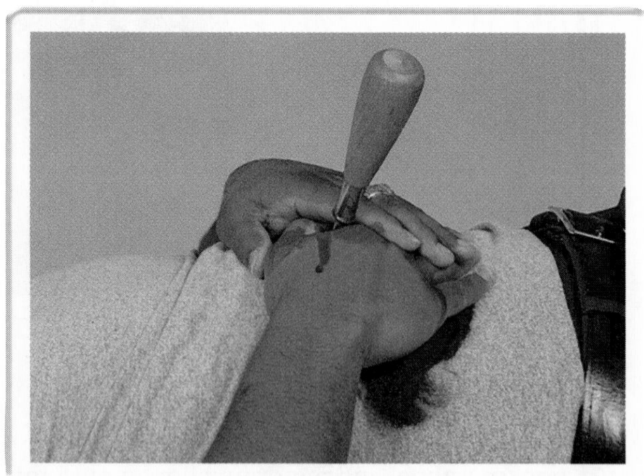

Distal circulation, sensation, and motor function should be assessed if the object is penetrating an extremity, shoulder, upper torso, groin, or hip.

continued

Using aseptic technique, apply sterile dressings over the wound and around the impaled object. Stabilize the object by applying bulk dressings on all four sides.

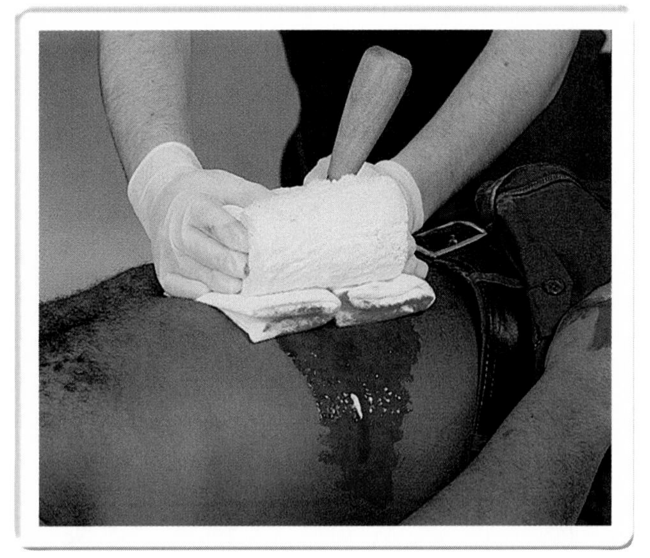

Bandage in place using sufficient pressure and bulk to control bleeding and stabilize the object without impeding circulation.

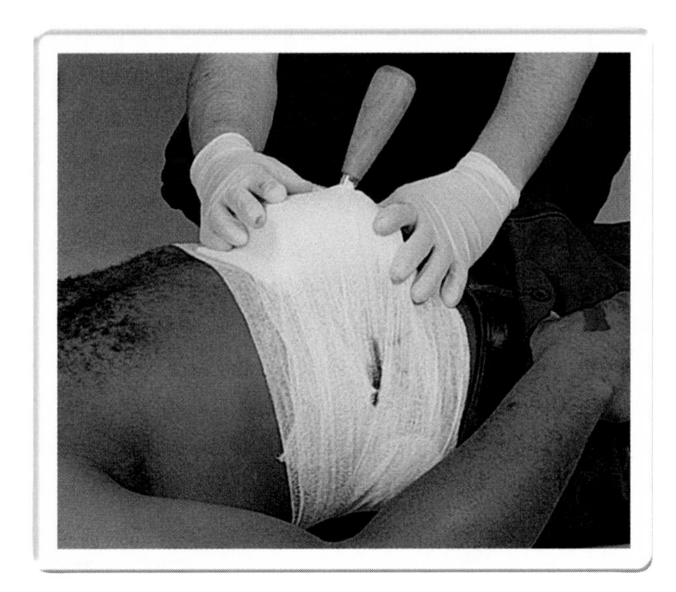

Reassess circulation, sensation, and motor function as appropriate.

continued

Amputations

Assessment of distal circulation, sensation, and motor function is impossible.

Using aseptic technique, apply a dry, sterile dressing to the stump of the extremity. Apply additional dressing materials as necessary to control hemorrhage. On rare instances, it may be necessary to apply a tourniquet above the injury to control bleeding. In most cases, however, direct pressure will be sufficient.

Anchor the roller bandage by wrapping a few turns around the extremity about 2 to 3 inches above the amputation. Using recurrent turns, completely cover the stump and all dressing materials. Secure by carefully tucking the end of the roller bandage.

Rinse the severed end with sterile saline and blot dry with sterile dressing materials. Cover the severed end with sterile dressings and wrap the entire part to keep clean. Place the wrapped part in a plastic, watertight bag, and place on a bed of ice. Be sure to prevent the severed part from freezing. Transport the severed part with the patient to the hospital.

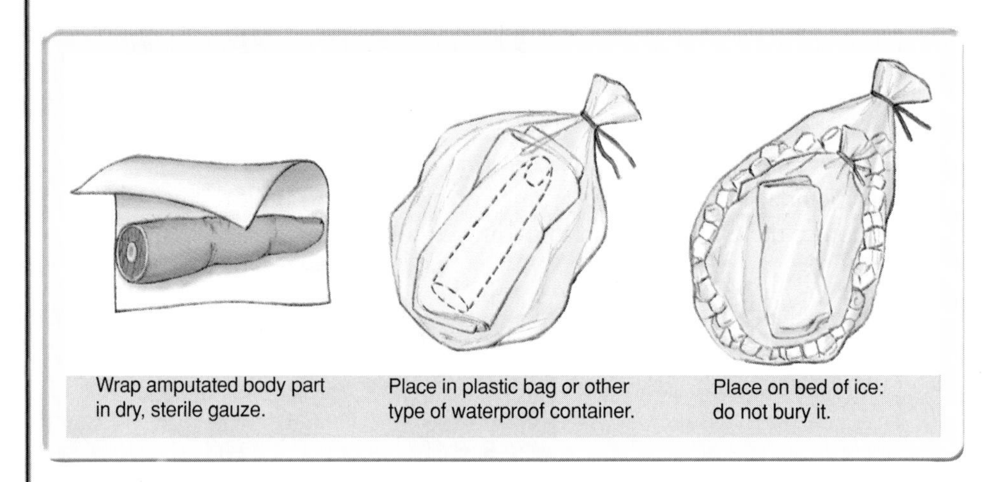

Wrap amputated body part in dry, sterile gauze. | Place in plastic bag or other type of waterproof container. | Place on bed of ice: do not bury it.

Burns

Smaller burns (less than 5% of the total body surface area) should be wrapped with dry dressings.

Cover all burned areas with sterile dressings using aseptic technique. Separate digits with gauze dressings unless fused together, and attempt to place them in a position of function. Gently wrap with loose, bulky roller bandage. Sling and swath upper extremity burns with the hand above the elbow.

Assess distal circulation often, if possible. Be prepared to loosen bandage and reapply if circulation becomes compromised due to swelling.

If a burn involves a large percentage of the body, wrap the victim in a dry burn sheet. No additional bandaging is necessary for severe burns.

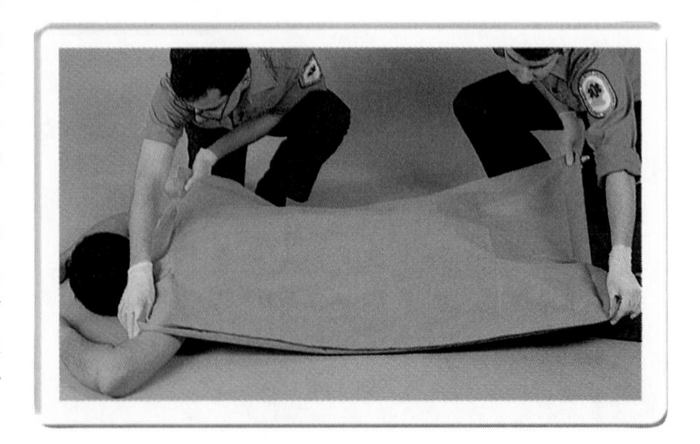

continued

Abdominal Organ Evisceration

Because of the injury location, no evaluation of distal circulation, sensation, or motor function is required.

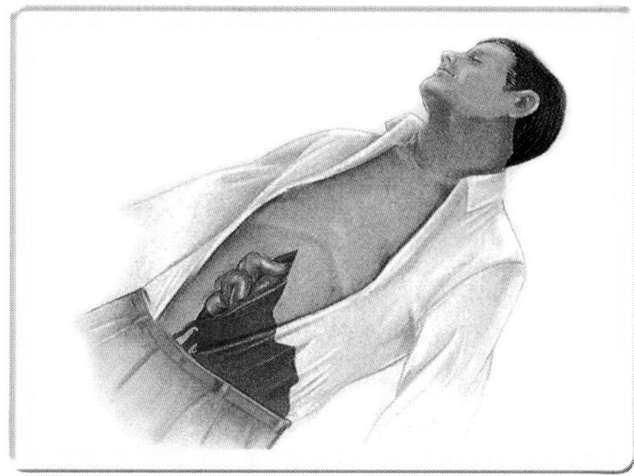

All protruding abdominal organs should be covered with a moist, sterile dressing using aseptic technique. Cover the moist dressing with an occlusive material (plastic package from multitrauma dressing, aluminum foil, etc.) to prevent wicking of contaminants. Do not attempt to replace the organs into the abdominal compartment.

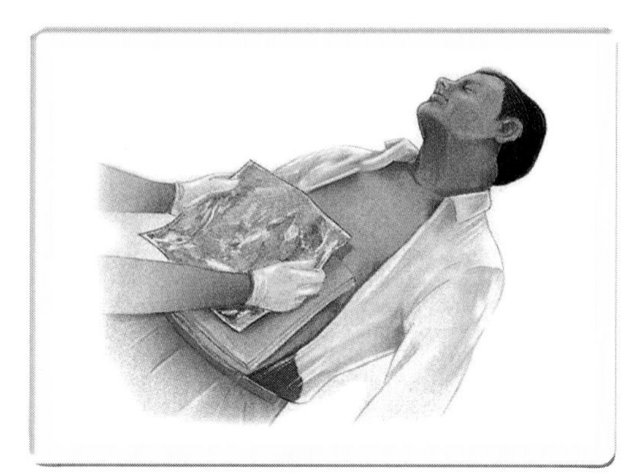

Secure the dressing in place using bandages or tape. Keep the eviscerated area and the patient warm.

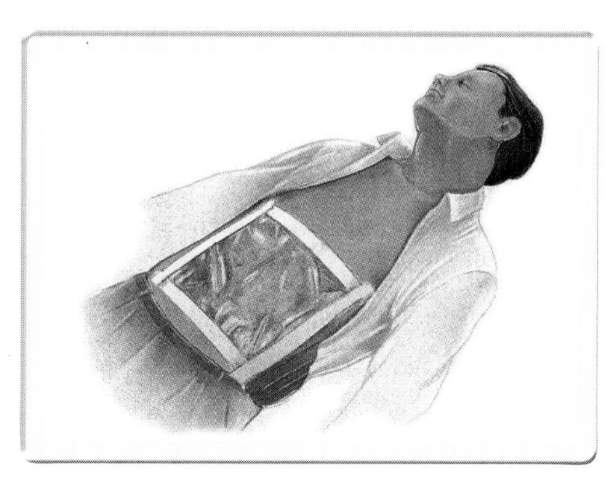

Reassess circulation, sensation, and motor function upon completion as required.

As appropriate and depending on the location of the injury, reassess circulation, sensation, and motor function as required upon completion.

SKILL 30

Joint Injury

Performance Objective

Given a patient and a description of the patient's injuries, the candidate(s) shall apply splinting techniques according to the principles of joint immobilization, in 10 minutes or less.

Equipment

The following equipment is required to perform this skill:

- Appropriate body substance isolation/personal protective equipment
- 3" roller bandages
- 6" roller bandages
- Triangular bandages
- 15" padded board splints
- 36" padded board splints
- 54" padded board splints

Equipment that may be helpful:

- Commercial splinting kits
- Vacuum splint kits
- Ladder splints
- Long backboards
- Backboard straps

Indications

- Suspected fracture or dislocation of a joint

Contraindications

- None

Complications

- None

Procedures

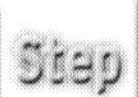

Step 1 Ensure body substance isolation before beginning procedures.

Prior to beginning patient care, appropriate body substance isolation procedures should be employed.

Step 2 Direct application of manual stabilization of the injury.

Upon identifying a joint injury, ensure the joint is stabilized manually. A second rescuer should gently hold the injury while the first rescuer assesses the injury and prepares for the immobilization.

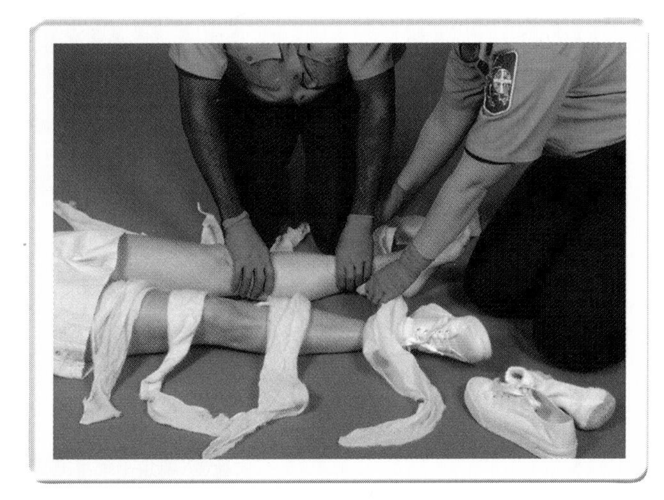

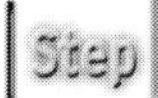 

Step 3 Assess distal circulation, sensation, and motor function (see Skill 27).

A check of the distal circulation through capillary refill or pulse check must be performed when appropriate and possible. Assessment of sensation and motor function must be performed on all extremities.

Joint injuries found without a pulse, sensation, or motor function should not be manipulated without approval from medical control. Manipulation of joint injuries can create devastating neurovascular conditions.

Step 4 Select proper splinting material.

Based on the specific injury, gather and prepare the appropriate splinting materials. Be sure enough material is available to perform the entire procedure without stopping.

See Skill 32 for specific immobilization techniques for various types of injuries.

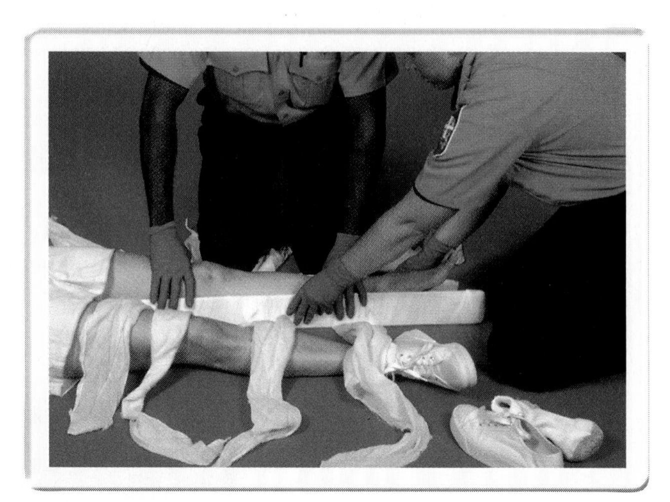

Step 5 Secure and stabilize the bone above the injury site.

Apply and secure splinting material to the bone above the injury.

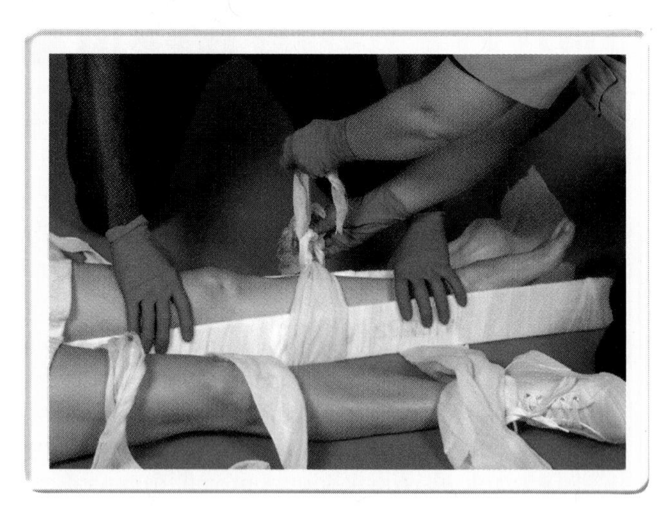

Step 6 Secure and stabilize the bone below the injury site.

Apply and secure splinting material to the bone below the injury.

Step 7 Secure the entire injured extremity.

After completing the application of the initial splinting material, confirm that effective motion restriction has been achieved. Ensure that the injury does not bear distal weight.

Step 8 Reassess distal circulation, sensation, and motor function.

A check of the distal circulation through capillary refill or pulse check must be performed when appropriate and possible. Assessment of sensation and motor function must be performed on all extremities.

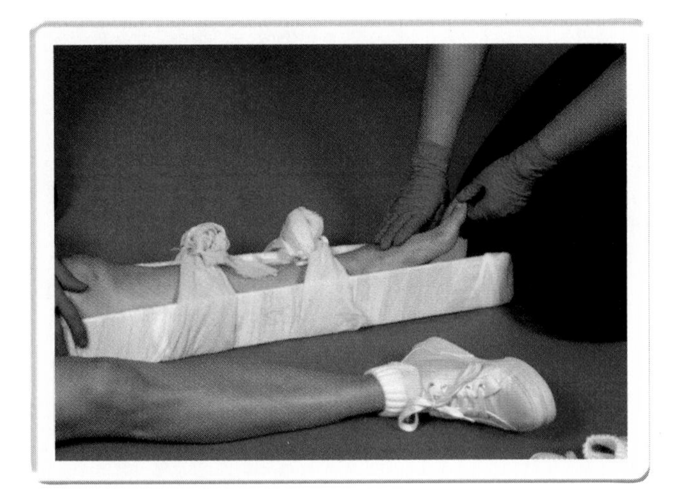

Step 9 Begin transport.

Patient should be readied for transportation to the emergency department. Simple joint injuries may be transported to non-trauma centers.

Step 10 Apply pain management.

Begin appropriate pain management. In many cases, ice or cold packs applied to the injury may be all that is available. In advanced life support systems, narcotic analgesics or inhaled anesthetics should be used.

SKILL 31 Long Bone Injury

Performance Objective

Given a patient and a description of the patient's injuries, the candidate shall apply splinting techniques according to the principles of fracture immobilization, in 10 minutes or less.

Equipment

The following equipment is required to perform this skill:

- Appropriate body substance isolation/personal protective equipment
- 3″ roller bandages
- 6″ roller bandages
- Triangular bandages
- 15″ padded board splints
- 36″ padded board splints
- 54″ padded board splints

Equipment that may be helpful:

- Commercial splinting kits
- Vacuum splint kits
- Ladder splints
- Long backboards
- Backboard straps

Indications

- Suspected or confirmed fractures to long bones
- Muscular or soft-tissue injury to arms or legs

Contraindications

- None

Complications

- None

Procedures

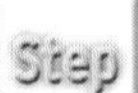 1

Ensure body substance isolation before beginning procedures.

Prior to beginning patient care, appropriate body substance isolation procedures should be employed.

Direct application of manual stabilization.

Upon identifying a long bone injury, ensure the joint is stabilized manually. A second rescuer should gently hold the injury while the first rescuer assesses the injury and prepares for the immobilization.

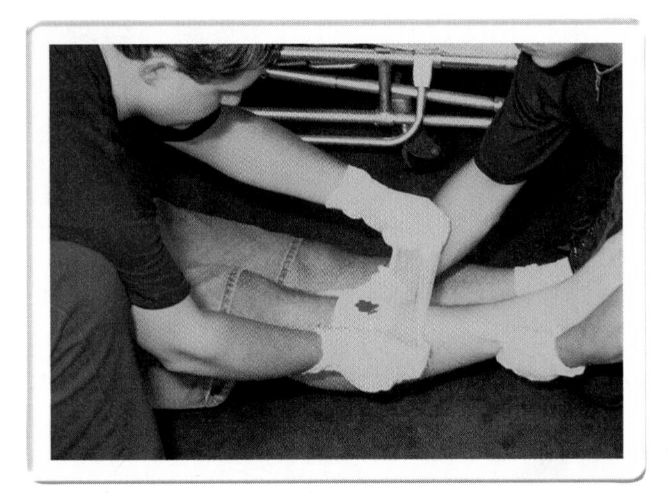

Assess distal circulation, sensation, and motor function (see Skill 27).

A check of the distal circulation through capillary refill or pulse check must be performed when appropriate and possible. Assessment of sensation and motor function must be performed on all extremities.

Angulated long bone injuries found without a pulse, sensation, or motor function should be slightly straightened with gentle traction in an attempt to return a blood flow into the extremity.

Select proper splinting material and measure splint.

Based on the specific injury, gather and prepare the appropriate splinting materials. Be sure enough material is available to perform the entire procedure without stopping.

Choose board splints or a commercially made device of appropriate size for the patient and the injury. Gently and without unnecessary movement, place the splint on the injured extremity.

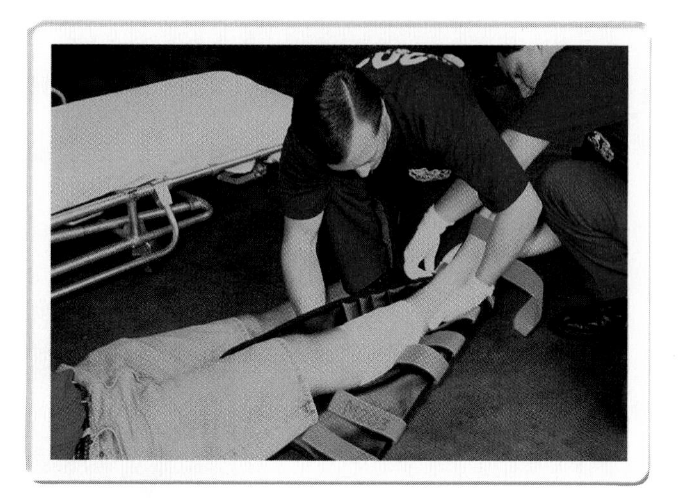

See Skill 32 for specific immobilization techniques for various types of injuries.

Secure and stabilize the joint above the injury site.

Apply and secure splinting material to the joint above the injury.

Secure and stabilize the joint below the injury site.

Apply and secure splinting material to the joint below the injury.

Secure the entire injured extremity.

After completing the application of the initial splinting material, confirm that effective motion restriction has been achieved. Ensure that the injury does not bear distal weight.

Step 8 Ensure position of function in hand or foot.

Before completion of the splinting process, ensure that the involved hand or foot is placed in proper position of function.

Step 9 Reassess distal circulation, sensation, and motor function.

A check of the distal circulation through capillary refill or pulse check must be performed when appropriate and possible. Assessment of sensation and motor function must be performed on all extremities.

Step 10 Begin transport.

Patient should be readied for transportation to the emergency department. Simple long bone injuries may be transported to non-trauma centers.

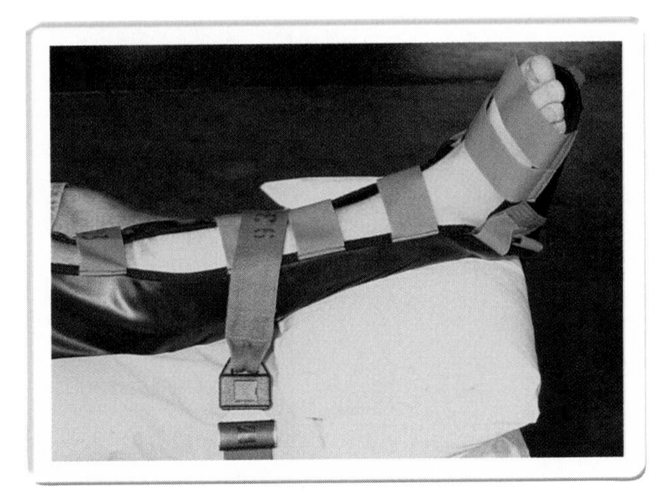

Step 11 Apply pain management.

Begin appropriate pain management. In many cases, ice or cold packs applied to the injury may be all that is available. In advanced life support systems, narcotic analgesics or inhaled anesthetics should be used.

SKILL 32

Suggested Procedures for Immobilization

Performance Objective

Candidates should be able to perform the following immobilization techniques. Alterations to these techniques are permitted provided the objectives of the skill are met.

Equipment

The following equipment is required to perform this skill:

- Appropriate body substance isolation/personal protective equipment
- 3″ roller bandages
- 6″ roller bandages
- Triangular bandages
- 15″ padded board splints
- 36″ padded board splints
- 54″ padded board splints

Equipment that may be helpful:

- Commercial splinting kits
- Vacuum splint kits
- Ladder splints
- Long backboards
- Backboard straps (minimum of three)

Indications

- None

Contraindications

- None

Complications

- None

Procedures

Step 1 Ensure body substance isolation before beginning procedures.

Prior to beginning patient care, appropriate body substance isolation procedures should be employed.

Step 2 Evaluate distal circulation, sensation, and motor function as required.

As appropriate and depending on the location of the injury, assess circulation, sensation, and motor function distal to the injury site.

Step 3 Apply appropriate immobilization specific to the injury.

Immobilize the injury. Procedures specific to common injuries are detailed in the following subsections.

Shoulder Injury

Assess circulation, sensation, and motor function distal to the injury.

Maintain the injury in the position found. Do not attempt to reposition the arm next to the body. Pillows or other forms of bulk may be needed to maintain the position of the arm. Apply a modified wrist sling around the opposite side of the neck. Secure with a minimum of two swathes. One of the two swathes should be positioned to prevent wrist drop.

Reassess circulation, sensation, and motor function.

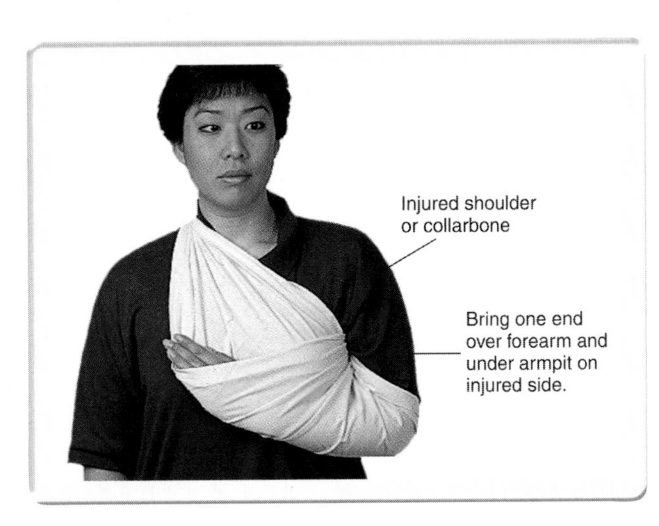

Step 3 continued

Scapula Fracture

Assess circulation, sensation, and motor function distal to the injury.

Maintain the injury in the position found. Do not attempt to reposition the arm next to the body. Apply a wrist sling around the neck. Secure with two swathes. One of the two swathes should be positioned to prevent wrist drop (as shown in the "Shoulder Injury" section).

Reassess circulation, sensation, and motor function.

Clavicle Injury

Assess circulation, sensation, and motor function distal to the injury.

Apply a modified wrist sling around the opposite side of the neck.

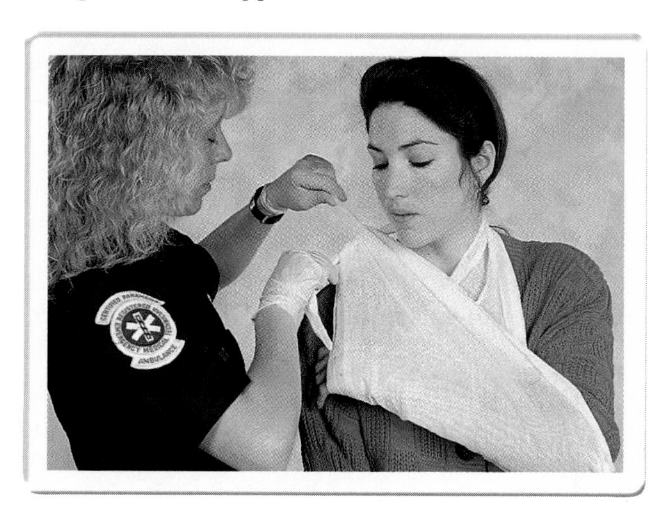

Secure with two swathes. One of the two swathes should be positioned to prevent wrist drop (see the photo shown in the "Shoulder Injury" section).

Reassess circulation, sensation, and motor function.

Humerus Fracture

Assess circulation, sensation, and motor function distal to the injury.

Most fractured humeri are found with the arm next to the body. Generally a board splint is not needed to immobilize this injury.

However, if the arm is away from the body and cannot be gently brought into anatomic position without excessive pain to the patient, use a short arm board applied laterally. Apply a modified wrist sling around the opposite side of the neck. Secure with two swathes. One of the two swathes should be positioned to prevent wrist drop.

Reassess circulation, sensation, and motor function.

continued

Elbow Injury: Joint Flexed, Arm Held Against Body

Assess circulation, sensation, and motor function distal to the injury.

Maintain the injury in the position found. No board splint is needed for this injury. Apply a wrist sling or modified wrist sling around the neck.

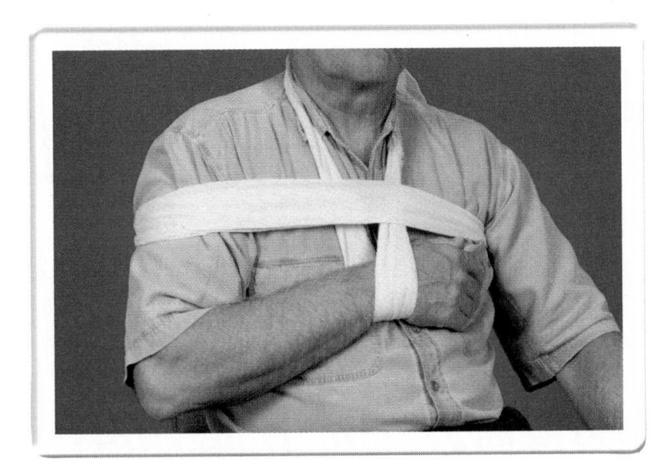

Secure with two swathes. One of the two swathes should be positioned to prevent wrist drop.

Reassess circulation, sensation, and motor function.

Elbow Injury: Joint Extended

Assess circulation, sensation, and motor function distal to the injury.

Maintain the injury in the position found. Apply a rigid splint from the upper arm to the fingertips. The board can be placed medial, lateral, or posterior depending on needs. Secure the splint with triangular bandages above and below the elbow, and above and below the wrist.

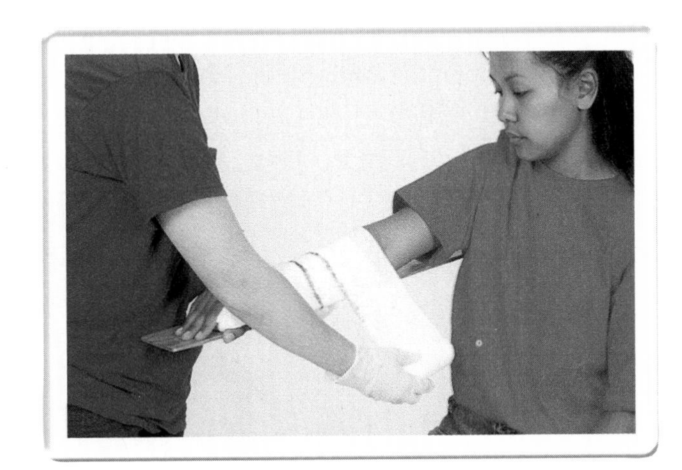

continued

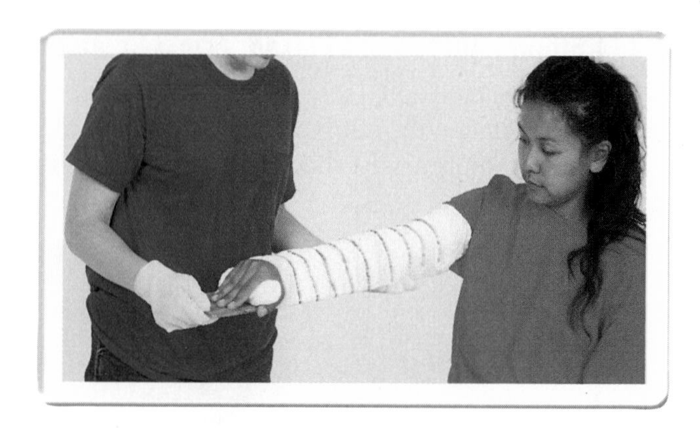

Elevate the injury with a pillow once the patient is supine.

Reassess circulation, sensation, and motor function.

Elbow Injury: Joint Flexed, Arm Away from Body

Assess circulation, sensation, and motor function distal to the injury.

Maintain the injury in the position found. Apply a rigid splint diagonal to the injury.

Secure the board to the humerus, below the elbow, and above and below the wrist. Either roller bandage or a triangular bandage may be used. Selection should be made depending on which can best secure the board with the least amount of movement. Roller bandage will work best most of the time. Keep the hand in a position of function if possible.

Move the arm into the body and apply a wrist sling around the neck. Secure with two swathes. One of the two swathes should be positioned to prevent wrist drop.

Reassess circulation, sensation, and motor function.

Radius or Ulna Fracture

Assess circulation, sensation, and motor function distal to the injury.

Position the arm with the elbow flexed, allowing the hand to rest higher than the elbow. Apply a rigid splint along the long axis of the arm from the elbow to the fingertips.

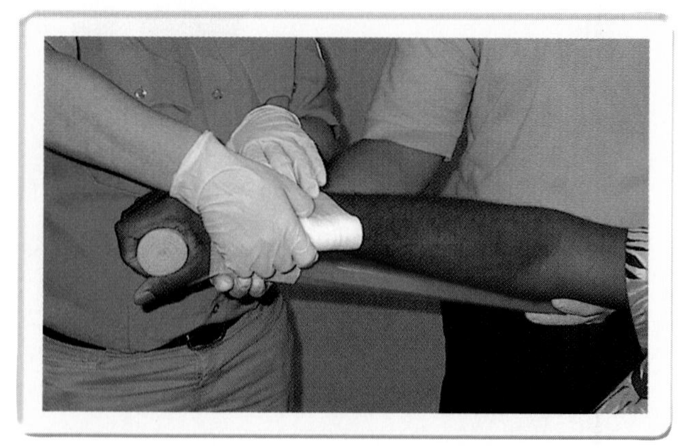

continued

Keeping the hand in a position of function, secure the board above and below the injury and above and below the wrist using roller bandage or triangular bandages. Selection should be made depending on which can best secure the board with the least amount of movement. Roller bandage will work best most of the time.

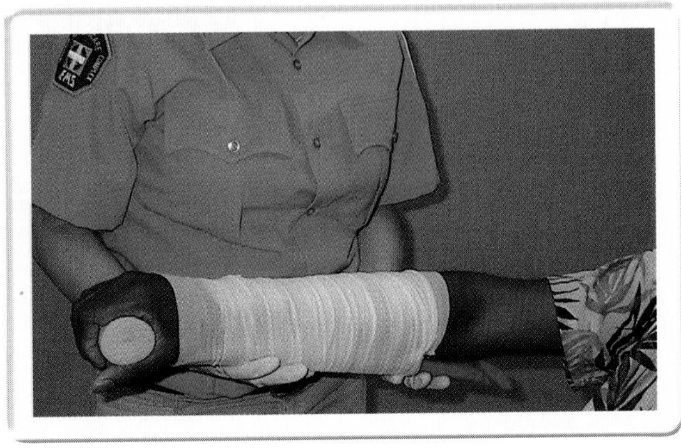

Move the arm into the body and apply a wrist sling around the neck. Secure with two swathes.

Reassess circulation, sensation, and motor function.

Wrist (Colles) Fracture

Assess circulation, sensation, and motor function distal to the injury.

Position the arm with the elbow flexed, allowing the hand to rest higher than the elbow.

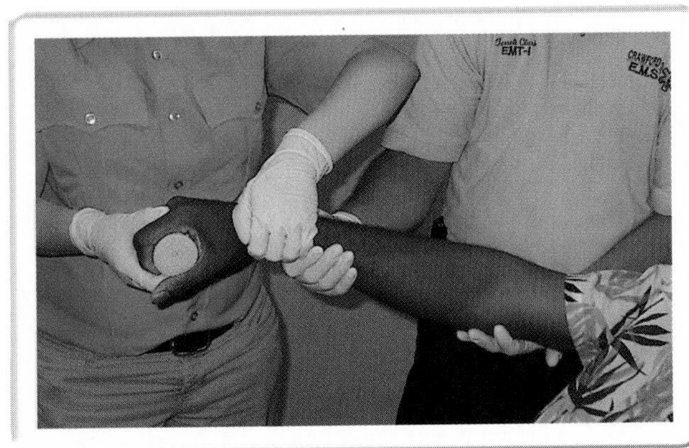

Apply a rigid splint along the long axis of the arm from the elbow to the fingers. Keeping the hand in a position of function, secure the board above and below the injury and above and below the wrist using roller bandage.

Step 3 continued

Support all gaps using contour padding made from dressing material or roller bandage.

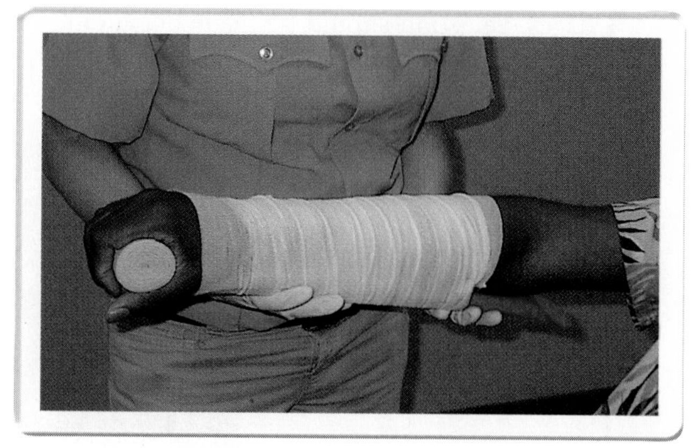

Move the arm into the body and apply a wrist sling around the neck. Secure with two swathes.

Reassess circulation, sensation, and motor function.

Femur Fracture: No Traction Splint Available or Compound Fracture

Assess circulation, sensation, and motor function distal to the injury.

Place a long board splint lateral to the leg, from the armpit past the ankle. A medium board splint should be applied medial to the leg from the groin past the ankle. Secure the boards above and below the knee, above and below the fracture site, and above and below the ankle. Position on a long backboard.

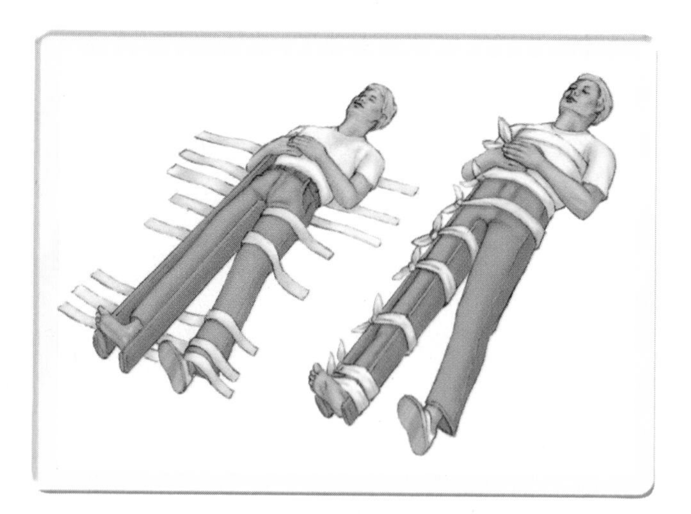

Reassess circulation, sensation, and motor function.

continued

Femur Fracture: Proximal Injury

Assess circulation, sensation, and motor function distal to the injury.

The leg should be splinted in a 90/90 technique. Using pillows or other appropriate materials, position the injured thigh at a 90° angle to the torso.

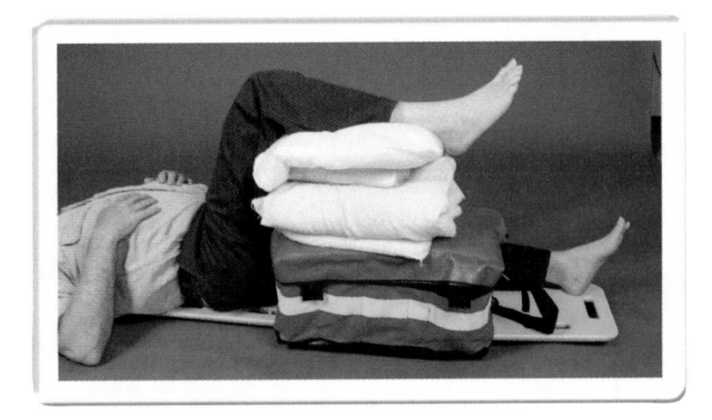

Secure the knee at a 90° angle to the thigh. Secure the entire leg to prevent movement during transport.

Reassess circulation, sensation, and motor function.

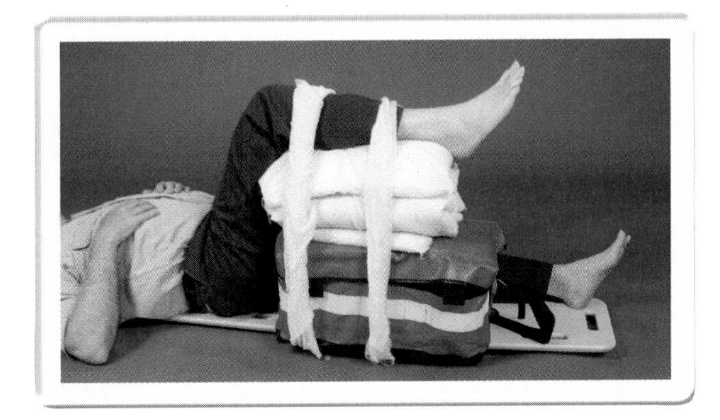

Hip or Pelvis Injury

Assess circulation, sensation, and motor function distal to the injury.

Place the patient on a long backboard. Stabilize the legs and hip in position found using pillows, blankets, or sheets. Secure with straps or triangular bandages.

Reassess circulation, sensation, and motor function.

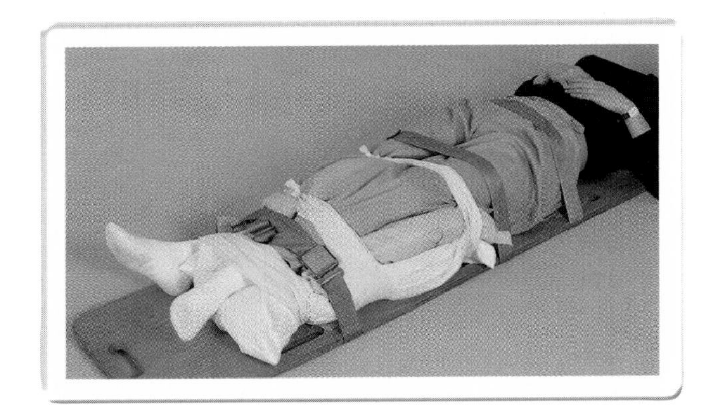

Knee Injury

Assess circulation, sensation, and motor function distal to the injury.

If the knee is flexed, splint the leg in the position found.

Place medium board splints on the lateral and medial sides of the leg. Secure using triangular bandages above and below the knee, and above and below the ankle.

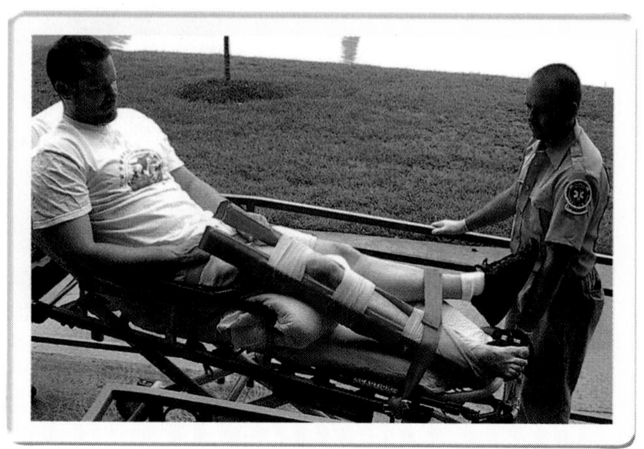

If the knee is straight, splint it straight.

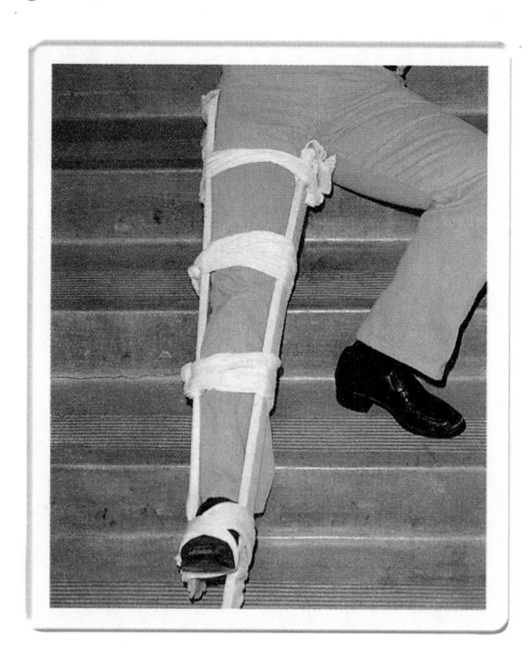

Reassess circulation, sensation, and motor function.

 continued

Tibia or Fibula Fracture

Assess circulation, sensation, and motor function distal to the injury.

Place medium board splints on the lateral and medial sides of the leg. Secure using triangular bandages above and below the knee, and above and below the ankle.

Reassess circulation, sensation, and motor function.

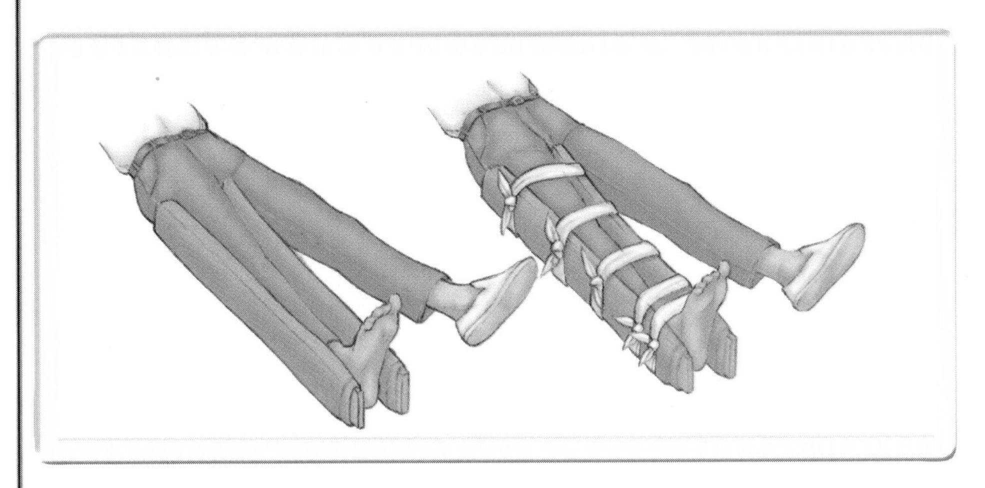

Ankle or Foot Injury

Assess circulation, sensation, and motor function distal to the injury.

Place medium board splints on the lateral and medial sides of the leg. Secure using triangular bandages above and below the ankle. A pillow may be used if it is long enough to immobilize the ankle completely.

Reassess circulation, sensation, and motor function.

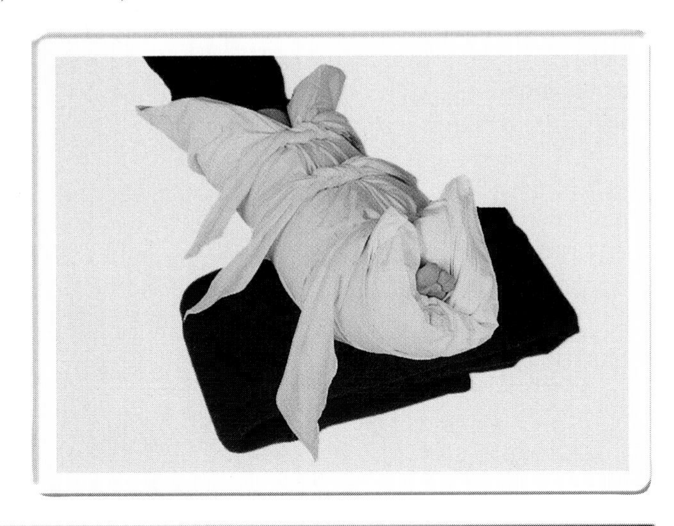

 4 **Reassess circulation, sensation, and motor function upon completion as required.**

Reassess circulation, sensation, and motor function as required upon completion.

SKILL 33

Spinal Motion Restriction—Seated

Performance Objective

Given a patient with a potential spinal injury, the candidates shall use the proper technique to immobilize the spine using short and long spinal immobilization devices, in 10 minutes or less.

Equipment

The following equipment is required to perform this skill:

- Appropriate body substance isolation/personal protective equipment
- Vest-style immobilization device, with
  - Neck pad
  - Chest and ischial straps
  - Head and chin straps
- Rigid extrication collars (various sizes or adjustable)
- Long backboard
- Backboard straps (minimum of three)
- Pillows

Equipment that may be helpful:

- Towels (for towel rolls and padding)
- Triangular bandages
- 3" roller bandages
- Web-type backboard straps

Indications

- Suspected or confirmed injury to the spine of a seated patient

Contraindications

- Multisystem trauma requiring rapid transport
- Environmental or situational hazards in which rapid removal is indicated

Complications

- Chest straps that are too tight may impede respiratory effort.

Procedures

Step 1 Ensure body substance isolation before beginning procedures.

Prior to beginning patient care, appropriate body substance isolation procedures should be employed.

Step 2 Direct assistant to place or maintain patient's head in the neutral, in-line position.

Have your partner take control of the seated patient's head and neck. This *must* be performed as soon as patient access is achieved. Ensure that the head and neck are maintained in the neutral position until they are *completely* stabilized with a rigid extrication collar and mechanical cervical motion restriction.

Tell the patient what is happening. Talking to the patient, conscious or not, and explaining what is happening will add to the success of the procedure.

Step 3 Assess distal circulation, sensation, and motor function in each extremity.

Check the distal circulation, sensation, and motor function in all four of the patient's extremities.

Step 4 Apply an appropriately sized rigid extrication collar.

Using an appropriate sizing method, choose the correct-sized rigid extrication collar for the patient. Explain to the patient the procedure for applying—and the need for—a rigid extrication collar. Apply the extrication collar without allowing flexion or extension of the patient's neck.

Manual immobilization must not be released to apply the extrication collar.

Step 5 Position the immobilization device behind the patient.

Without compromising the integrity of the cervical spine, position the short device behind the patient. Begin by having your partner apply manual traction to the patient's head and neck while moving the patient's torso forward. All movement of the patient should be slow, steady, and coordinated. Usually a three count, given by the responder controlling the head, is used to ensure all responders work as a unit in the move.

Slide the immobilization device behind the patient and manipulate into position. While working the immobilization device behind the patient, be sure the device does not get caught on the patient's clothing or the car seat. Bring the side panels up into the axilla as high as possible without impeding circulation through the axillary artery.

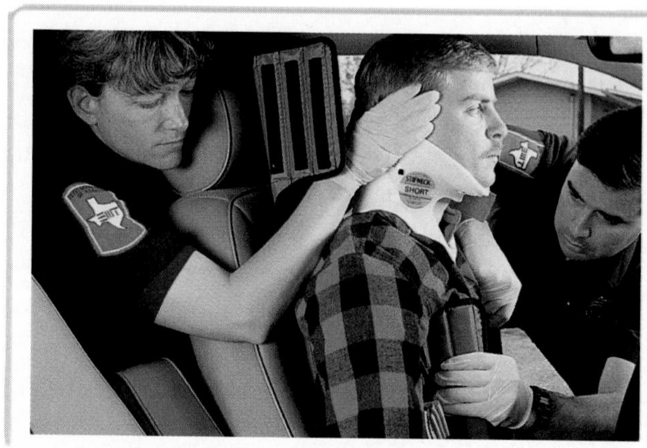

Safety Tips

The Standing Patient With Spinal Injury

Occasionally, you will encounter a walking patient following an incident with a high index of suspicion for spinal injury. In such situations it is best to have the patient remain standing and perform spinal immobilization in the upright position.

Step 6 Secure the device to the patient's torso.

Instruct your partner to move the patient back into position against the back of the seat. Begin strapping the short device to the patient's chest. Attach all straps smoothly and snugly. Be sure to adjust each strap without twisting the device. The chest strap should be snug but should not compromise the patient's respiratory effort. Attach both ischial straps. These straps may be crossed or secured directly.

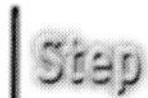 

Step 7 Evaluate torso fixation.

Before securing the head to the device, ensure the splint is firmly secured to the torso. The top strap should be securely tightened without impeding the patient's ventilatory effort. Adjust the device as necessary to ensure proper fit.

Step 8 Evaluate and pad behind the patient's head as necessary.

It may be necessary to place a folded neck pillow behind the patient prior to achieving neutral alignment with the back of the board. Inspect the void between the patient and the device to determine the amount of padding required. When completed, the patient's face should be positioned looking directly forward, in a natural anatomic position. Be sure that the patient's neck is not flexed or hyperextended.

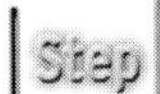

Step 9 Secure the patient's head to the device.

Bring the head panels around both sides of the head and secure them in place with foam straps applied to the forehead. Chin straps may be needed as well. Neutral alignment of the patient's neck must be maintained by the partner throughout this step.

Once this step is complete, tighten the chest strap. The head and neck must be in a neutral, in-line position when finished.

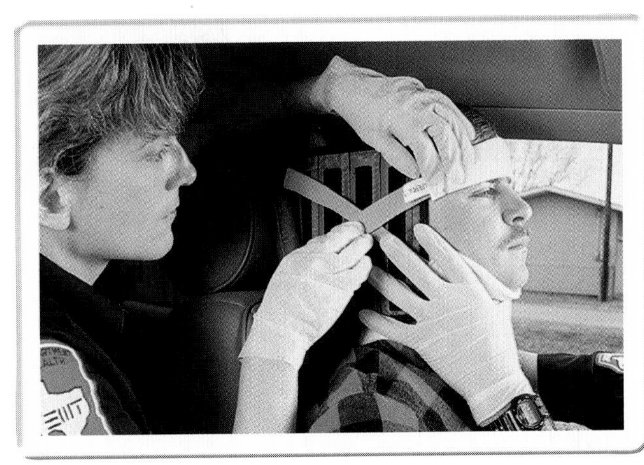

Step 10 Move the patient to a long backboard.

With the patient secured to the short device, move the patient to a long backboard.

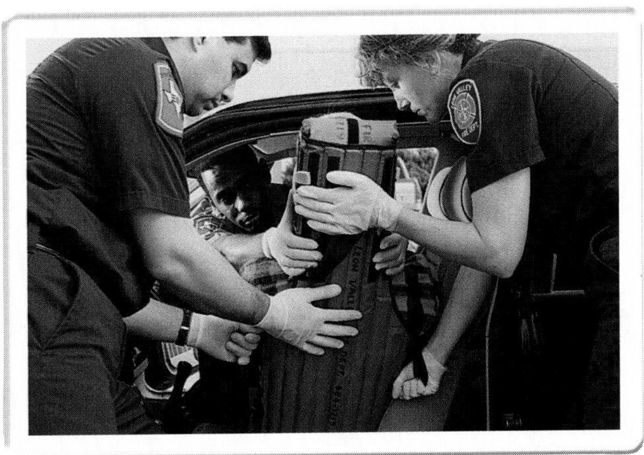

In the Field

Rapid Extrication

In situations in which the patient's condition or situation requires immediate removal from a vehicle, there is no time to apply a vest-style device. In these cases rapid extrication should be employed. This procedure begins identical to seated spinal motion restriction, but uses the rescuer's body as a support for the patient's torso as he or she is rotated to be laid onto a long backboard.

Follow these steps:

1. Support the cervical spine.
2. Apply a rigid extrication collar, sized appropriately for the patient.
3. Position the backboard on the cot or the seat next to the patient.
4. Carefully rotate the patient's upper and lower torso as a unit. Be sure the neck and legs follow appropriately.
5. Position the backboard under the patient's buttocks.
6. Lower the patient onto the backboard, maintaining neutral alignment of the spine.
7. Secure the patient to the backboard *before* removal from the vehicle and quickly assess pulse, sensation, and motor function in all four extremities.
8. Note in the patient care report the reason for the rapid extrication of the patient and the status of pulse, sensation, and motor function in all extremities before and after the move.

In the Field

Lifting and Carrying the Patient on a Long Backboard

It is always best to lift a backboard directly from the ground to the stretcher. When it is necessary to carry a patient on a backboard for any distance, the following techniques can be used.

- **Diamond carry.** This carry requires four rescuers. The two primary rescuers are positioned at the patient's head and feet, and use both hands to lift the patient. The remaining rescuers use one hand to carry the backboard from the center handholds. This carry is usually reserved for carrying the patient short distances.
- **One-handed carry.** This carry requires four or six rescuers. Each rescuer grasps a handhold at the chest or legs of the patient. If six rescuers are available, the additional two grab the center handholds. The patient is then lifted and carried to the desired location.

Step 10 continued

Place the end of the board under the patient's hip and rotate the patient *as a unit* in line with the long backboard. Carefully lower the patient onto the backboard.

Using longitudinal pulls, position the patient on the long backboard. Once the patient is properly positioned on the board, the ischial straps can be loosened. Strap and secure the patient to the long backboard. Remember that all patient movement must occur as a coordinated unit.

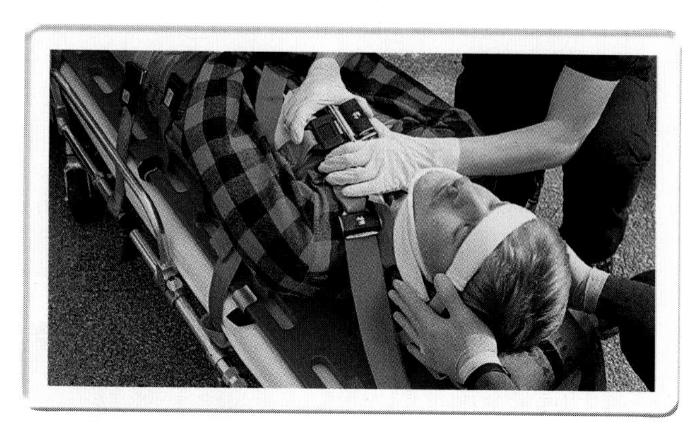

Step 11 Reassess distal circulation, sensation, and motor function in each extremity.

Check the distal circulation, sensation, and motor function in all four of the patient's extremities.

Spinal Motion Restriction—Supine

Performance Objective

Given a patient with a potential spinal injury, the candidates shall use the proper technique to immobilize the spine using short and long spinal immobilization devices, in 10 minutes or less.

Equipment

The following equipment is required to perform this skill:

- Appropriate body substance isolation/personal protective equipment
- Rigid extrication collars (various sizes or adjustable)
- Long backboard
- Backboard straps (minimum of three)
- Cervical immobilization device
- Pillows

Equipment that may be helpful:

- Towels (for towel rolls and padding)
- Web-type backboard straps
- Triangular bandages
- Adhesive tape

Indications

- Suspected or confirmed spinal injury

Contraindications

- None

Complications

- Decubitus ulcers
- Back pain

Procedures

Step 1 Ensure body substance isolation before beginning procedures.

Prior to beginning patient care, appropriate body substance isolation procedures should be employed.

Step 2 Direct assistant to place or maintain patient's head in the neutral, in-line position.

Have your partner take control of the patient's head and neck. This *must* be performed as soon as patient access is achieved. Ensure the head and neck are maintained in the neutral position until they are *completely* stabilized with a rigid extrication collar and mechanical cervical motion restriction.

Tell the patient what is happening. Talking to the patient, conscious or not, and explaining what is happening will add to the success of the procedure.

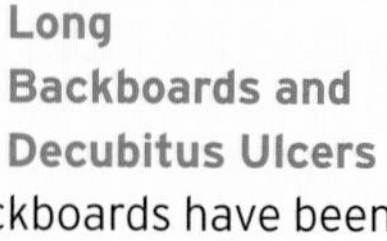

Safety Tips

Long Backboards and Decubitus Ulcers

Long backboards have been shown to greatly accelerate the incidence of decubitus ulcers in patients. Consider padding the entire length of the backboard, or using a scoop stretcher in lieu of a standard backboard. Scoop stretchers can be applied in an identical manner as described in this skill and simply broken apart upon arrival at the hospital.

Step 3 Assess distal circulation, sensation, and motor function in each extremity.

Check the distal circulation, sensation, and motor function in all four of the patient's extremities.

Step 4 Apply an appropriately sized rigid extrication collar.

Using an appropriate sizing method, choose the correct-sized rigid extrication collar for the patient. Explain to the patient the procedure for applying—and the need for—a rigid extrication collar. Apply the extrication collar without allowing flexion or extension of the patient's neck.

Manual immobilization must not be released to apply the extrication collar.

Step 5 Position the long backboard appropriately.

Position the long backboard beside the patient.

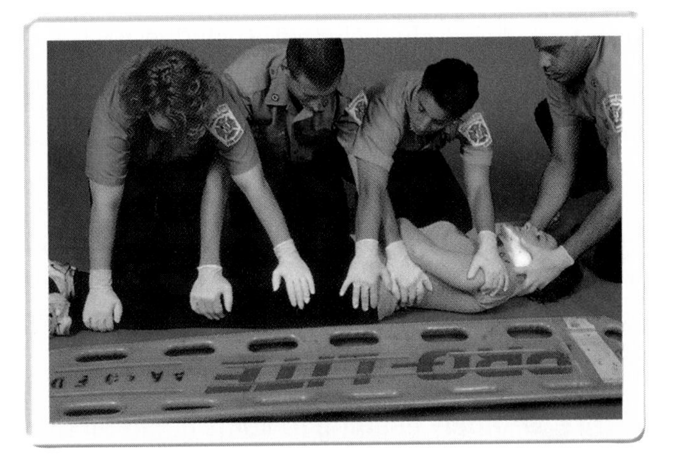

Step 6 Direct movement of the patient onto the long backboard without compromising the integrity of the spine.

With the aid of one or two *additional* rescuers, perform a log roll of the patient. The log roll should be performed by grasping the patient's torso and pelvis. With direction from the person stabilizing the head, roll the patient as a unit. A proper log roll should result in the entire spine being lifted without twisting or shifting.

In the Field

Rolling Patients

Log rolls and other procedures used to move a patient require a firm grasp of the patient, not the patient's clothing. Avoid attempting to move a patient by grabbing a collar, sleeve, or belt loop. Waistbands of pants are acceptable holding points only because they are usually sturdier than other pieces of clothing.

Also remember that if the patient is to be moved as a unit, all handholds should be made on the patient's axial skeleton, not the appendicular. This means that the torso should be grasped. Similarly, when controlling the head, a hold that incorporates the chest, neck, and head is much better than holding the head alone.

Step 6 continued

Perform a quick and thorough examination of the patient's back and spine. Place the long backboard directly under the patient and roll the patient down onto the board. Using small, controlled movements, slide or lift the patient into proper position on the board. The patient should be positioned by sliding up or down along the long axis of the body, using a pulling motion. Pushing the torso up or over is never appropriate and may result in compression or separation of the spine. Each movement of the patient should be a coordinated event. In most cases the responder at the patient's head has the responsibility for signaling the other responders, usually with a three count, to ensure that all responders act as a unit.

Step 7 Apply padding to voids between the torso and the backboard as necessary.

Place padding in any voids found between the torso and the spine to reduce the possibility of spinal movement.

Step 8 Secure the patient's torso to the backboard.

Secure the patient to the board by placing a minimum of three straps. The first strap should be over the patient's chest, the second over the pelvis, and the third over the thighs. If the handholds and strapping holes on the backboard are not aligned to properly position the straps over the chest or abdomen, crossing the straps may be necessary. This is accomplished by running straps from a higher and a lower hole position on each side of the patient and crossing them as they are placed on the opposite side.

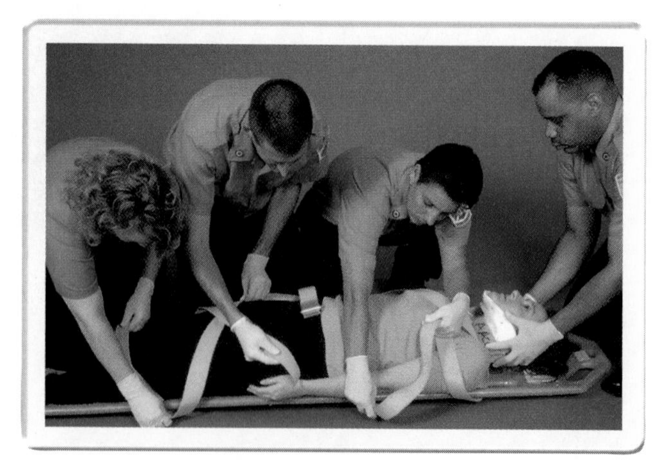

In the Field

Thinking Off the Board

Long backboards have become the mainstay for trauma management. It has become customary for every trauma patient, with or without spinal injury, to be placed on a long backboard for transport. Although a safe move in many situations, this can be detrimental in others. One such case would be a patient with a diaphragmatic tear. These injuries can allow the abdominal organs to enter the thorax if the patient is laid down, causing difficulty breathing and agitation. In such cases the patient should be allowed to sit up as long as the blood pressure is not compromised. If spinal injury is suspected along with the torn diaphragm, support can be accomplished with a vest-type immobilization device. Always evaluate the need for a long backboard and spinal motion restriction before application. As with all aspects of medicine, these decisions should be based on the patient's best interest rather than reflex.

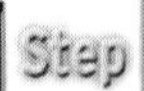 9

Evaluate and pad behind the patient's head as necessary to maintain neutral, in-line position.

Evaluate the need for padding behind the head to ensure neutral position.

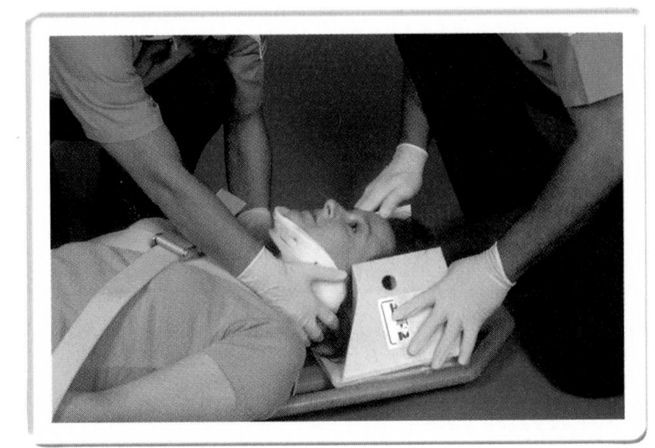

Most commercial cervical immobilization devices (CIDs) have a small pad as a part of the device. *Do not* assume that the padding on the commercially made device is sufficient padding. When completed, the patient's face should be looking skyward in the supine position, parallel to the long backboard.

Secure the patient's head to the long backboard.

Secure the patient's head to the long backboard using a cervical immobilization device, towel rolls, or other soft padding that restricts lateral movement. When complete, the head and neck must be in neutral, in-line position.

Secure the patient's legs to the backboard.

Place a pillow under the patient's knees to maintain a slightly bent position. This will place the back in the correct anatomic position. Ensure the patient's legs are secured to the long backboard by applying a fourth strap below the patient's knees.

Safety Tips

Helmet Removal

Helmets protect the patient from injury during an accident, but may pose obstacles to proper patient assessment and management if left in place. The decision to remove a helmet really is two-fold: Does the helmet disrupt cervical alignment, or can the airway be maintained? Before removing a helmet, evaluate the helmet and the situation.

- **Motorcycle helmets.** In most cases, motorcycle helmets should be removed prior to securing patients to long backboards. This will require two rescuers to work together. The first rescuer reaches into the helmet and supports the patient's head and neck. The second rescuer grasps the helmet by the chin straps and, pulling the straps out and up, gently removes the helmet.
- **Football helmets.** Football helmets are usually not removed in the prehospital setting. Since they are worn in conjunction with shoulder pads, cervical spine alignment is usually not compromised. By removing the face mask, easily accomplished with a tool called a Trainer's Angel, you can assess and maintain the airway. Removal of a football helmet without removal of the shoulder pads will require considerable padding to achieve cervical alignment and is usually not considered the standard of care.

Secure the patient's arms to the backboard.

Secure the patient's arms to the long backboard to prevent the arms from flailing when the board is lifted.

Reassess distal circulation, sensation, and motor function in each extremity.

Check the distal circulation, sensation, and motor function in all four of the patient's extremities.

Traction Splinting—Ratchet Device

Performance Objective

Given a traction splint and a patient, the candidates shall apply the traction splint to immobilize a closed fracture to the midshaft of the femur, in 10 minutes or less.

Equipment

The following equipment is required to perform this skill:

- Appropriate body substance isolation/personal protective equipment
- Ratchet-type traction splint, with:
  - Straps
  - Ankle hitch
  - Adjustable stand
- Long backboard
- Long backboard straps

Equipment that may be helpful:

- Triangular bandages
- Pillow

Indications

- Closed midshaft femur fracture

Contraindications

- Fractures of the proximal femur, hip, or knee
- Fractures of the lower leg on the same side

Complications

- None

Procedures

Ratchet-type traction splints require two people for proper application. Both partners have an equal responsibility to properly apply the device. Once the initial assessment has identified the fractured femur and the decision to apply a traction splint has been made, rescuers should work together using the following steps, with Rescuer One applying the splint and Rescuer Two pulling traction.

In the Field

What Is Traction?

Traction is the use of surrounding muscles to stabilize a fracture. Stretching the muscles supports the ends of the broken bone or separated joint. To achieve traction, only slight tension needs to be applied and maintained. The goal here is simply to pull on the muscles, not to realign the bone ends. Realignment or rearticulation is known as *reduction* and is performed only after further evaluation determines the correct course of action.

Step 1 Ensure body substance isolation before beginning procedures.

Prior to beginning patient care, appropriate body substance isolation procedures should be employed.

Step 2 Rescuer One: Apply manual stabilization of the injured leg.

Upon recognition of a midshaft fracture to the femur, the first rescuer should stabilize the fracture by positioning his or her hands gently around the thigh. The objective is to reduce involuntary movement in the patient's leg muscles and to act as a shock absorber while the remainder of the leg is assessed.

Rescuer Two: Assemble and prepare splint.

Ensure the splint is adjusted to the proper size using the uninjured leg. The splint should be approximately 10 to 12 inches longer than the distance from the ischial tuberosity to the ankle. Prepare four straps and position them on the splint. It is best to position the straps toward the top of the splint and to pull the straps down into position as needed. Attempting to guess the proper position of the straps on the splint will usually lead to having to adjust straps up, working against the taper of the splint.

Release the traction ratchet and pull the strap out to its full length. Release and open the ischial strap.

Rescuer One: Stabilize lower leg.

Support the lower leg while the ankle hitch is applied.

Pressure should be enough that movement at the ankle does not reach the femur, but not so much that the lower leg is injured.

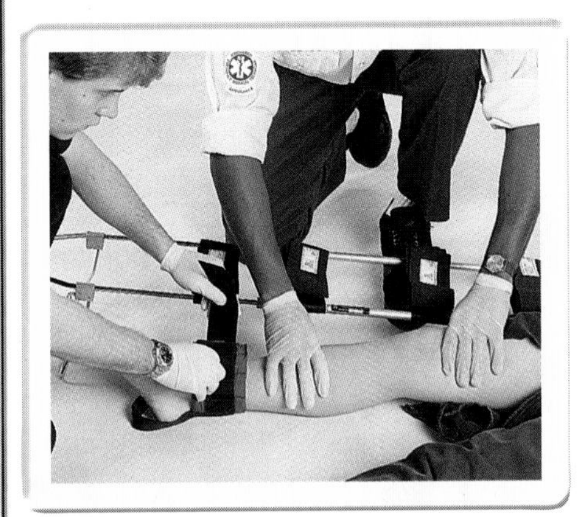

Rescuer Two: Assess distal circulation, sensation, and motor function. Apply ankle hitch.

If not already performed, carefully remove the patient's shoe.

Check the distal circulation through pulse check or capillary refill. Also assess the distal sensation and motor function.

With proper padding applied circumferentially around the ankle, apply an ankle hitch to the patient. The ankle hitch must be applied in a position to pull traction along the long axis of the bones of the lower leg. Minimal manipulation should be used to prevent additional injury to the fracture and surrounding soft tissue.

Rescuer One: Position hands to lift thigh.

Move hands up the leg and position under the thigh, one hand on each side of the fracture site.

Rescuer Two: Position hands to pull traction.

Support the limb as you position your hands to pull traction.

Step 5 Rescuer One: Lift thigh.

Give your partner the direction to lift the patient's leg and pull traction while you lift the thigh.

Maintaining proper body position will reduce the risk of back injury to the responder.

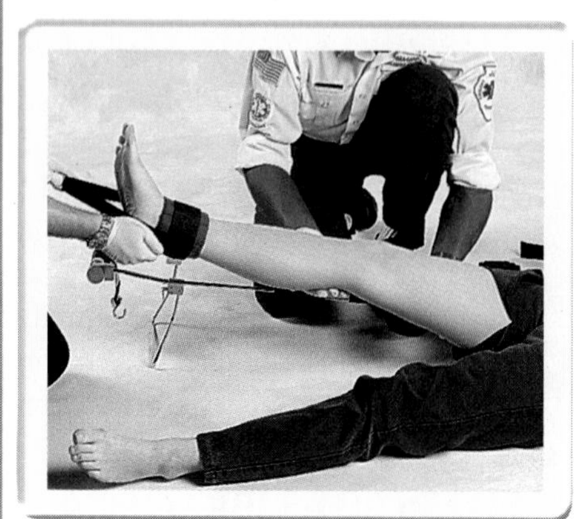

Rescuer Two: Lift leg and pull traction.

Working with your partner, gently lift the leg, pulling traction as you elevate. Sufficient elevation should be achieved to allow your partner to apply the splint under the leg. Once elevation and traction are achieved, the leg *must* remain in this position until mechanical traction is achieved and secured. Mechanical traction is secured after your partner secures the strap around the patient's ankle to the traction splint.

Maintaining proper body position will reduce the risk of back injury to the responder.

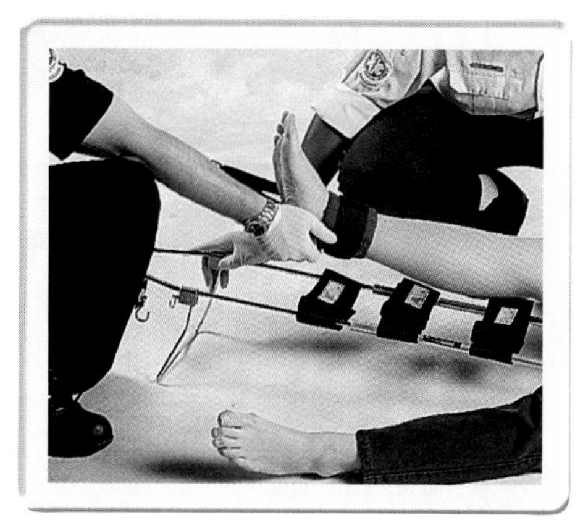

Step 6 Rescuer One: Apply splint.

Position the ischial pad under the patient's thigh and firmly against the ischium.

Apply padding across the inguinal region from both sides of the frame. Fasten the ischial strap tight enough to hold the splint, but not enough to impede circulation. Ensure the patient's genitalia are not trapped by the strap.

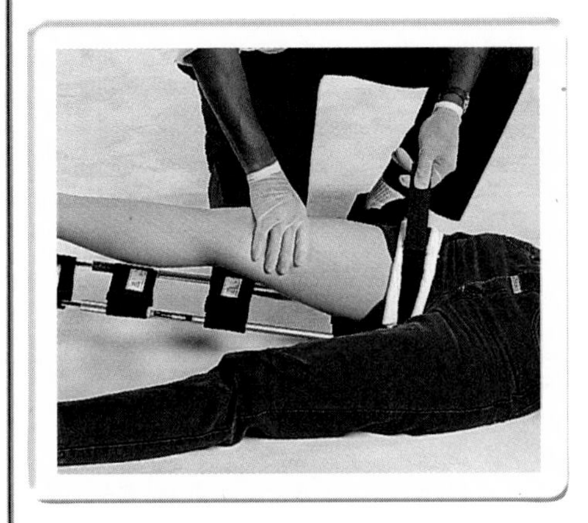

Rescuer Two: Pull traction.

Connect the ankle hitch to the frame and take up slack.

Using a mechanical ratchet or a windlass, tighten the connection of the ankle hitch. Once manual traction has been equaled, no further traction should be pulled (only 10% of the patient's weight is needed). If a windlass was used, secure the windlass to the frame to prevent traction from being lost.

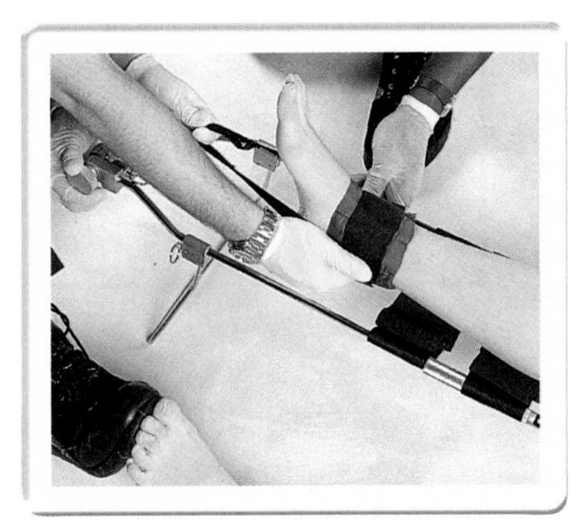

Step 7 Rescuer One: Secure ankle hitch.

Apply the splint strap just above the ankle. Mechanical traction is now complete.

Rescuer Two: Maintain control and elevation of the leg.

As mechanical traction is obtained, maintain elevation of the leg.

Rescuer One: Apply remaining straps and lower the stand.

Start by placing a strap below the ischial strap and continue down the leg, placing one strap above the fracture site, one strap below the fracture site, and the final straps below the knee. If cravats are being used, they should be tight enough to lift the leg into alignment with the ankle hitch. Cravats should be positioned to avoid placement directly over or around the fracture itself. However, Velcro straps can be placed over the fracture site if necessary. Knots should be tied to the splint and not over the patient's leg.

Lower the traction splint stand.

Rescuer Two: Lower splint and reassess patient.

Gently lower the splint to rest on the stand, and maintain control to prevent excessive movement.

Rescuer One: Reevaluate the proximal and distal securing devices.

Inspect the splint to ensure the device is working properly. Check the ischial strap (proximal securing device) to ensure it remains secure. The ankle hitch (distal securing device) should be checked to ensure that it remains snug and pulls straight along the long axis of the leg and does not pull from or on the foot.

Rescuer Two: Reassess distal circulation, sensation, and motor function.

Check the distal circulation through pulse check or capillary refill. Also assess the distal sensation and motor function.

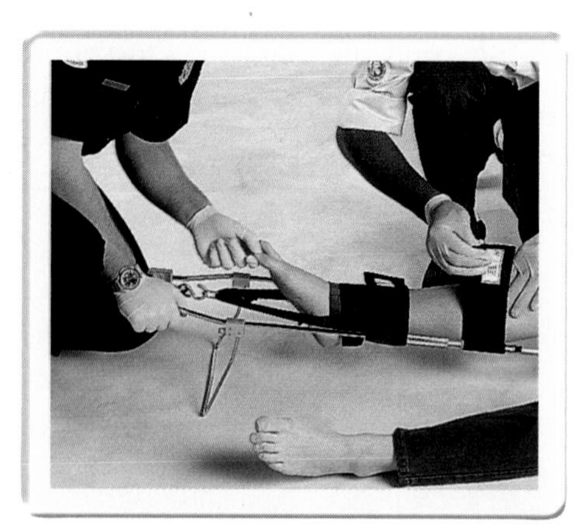

Step 10 Secure the torso and splint to the long backboard to immobilize the hip.

Once the traction splint has been applied, the patient should be moved onto a long backboard and secured so that the end of the traction splint is supported. Ensure that the stand remains on the long backboard so that the splint remains elevated.

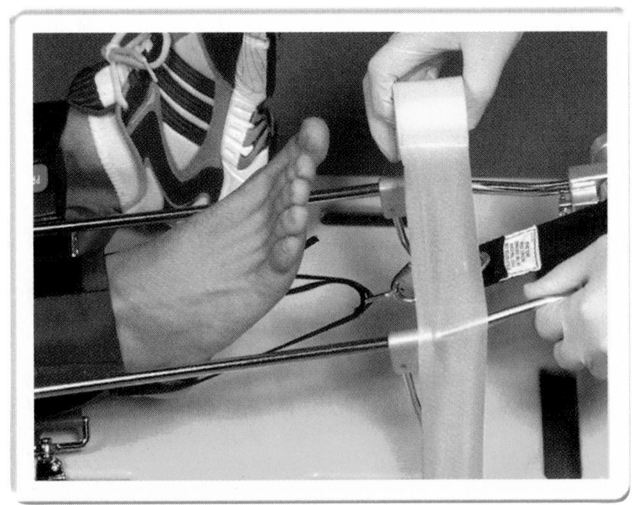

SKILL 36

Traction Splinting—Sager-Type Device

Performance Objective

Given a traction splint and a patient, the candidates shall apply the traction splint to immobilize a closed fracture to the midshaft of the femur, in 10 minutes or less.

Equipment

The following equipment is required to perform this skill:

- Appropriate body substance isolation/personal protective equipment
- Sager-type traction splint, with
  - Straps
  - Ankle hitch
- Long backboard
- Long backboard straps

Equipment that may be helpful:

- Triangular bandages
- Pillow

Indications

- Closed midshaft femur fracture

Contraindications

- Fractures of the proximal femur, hip, or knee
- Fractures of the lower leg on the same side

Complications

- None

Procedures

Step 1 Ensure body substance isolation before beginning procedures.

Prior to beginning patient care, appropriate body substance isolation procedures should be employed.

Step 2 Direct application of manual stabilization of the injured leg.

Upon recognition of a midshaft fracture to the femur, direct another rescuer to stabilize the fracture site.

The fracture site should be stabilized by positioning your hands gently around the thigh. The objective is to reduce involuntary movement in the patient's leg muscles and to act as a shock absorber while the remainder of the leg is assessed. During the application of the ankle hitch, move your hands down to hold the lower leg. Pressure should be enough that movement at the ankle does not reach the femur, but not so much that the lower leg is injured.

Step 3 Assess distal circulation, sensation, and motor function.

Check the distal circulation through pulse check or capillary refill. Also assess the distal sensation and motor function.

Step 4 Position the splint.

Position the splint between the patient's legs, resting the ischial perineal cushion (the saddle) against the ischial tuberosity, with the shortest end of the articulating base toward the ground. In the case of a unilateral fracture, the splint should be placed in the perineum on the side of the injury. In bilateral fractures, excluding pelvic trauma, the splint should be placed on the side with the greatest degree of injury.

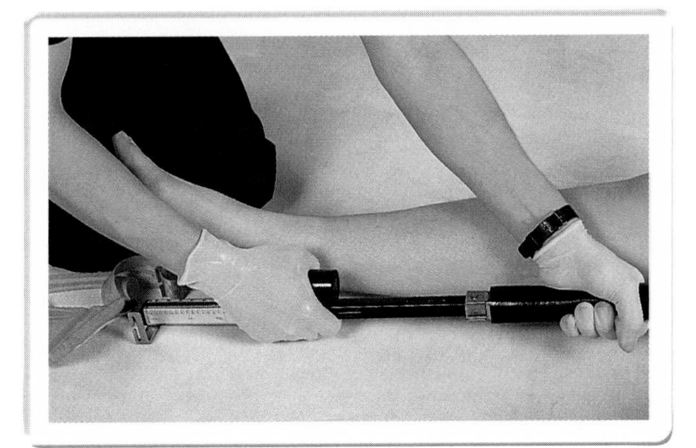

Step 5 Apply the thigh strap and adjust the splint.

Apply the thigh strap around the upper thigh of the fractured limb. Push the ischial perineal cushion gently down while at the same time pulling the thigh strap laterally against the ischial tuberosity. Tighten the thigh strap snugly. Lift the spring clip to extend the inner shaft of the splint until the crossbar rests adjacent to the patient's heels.

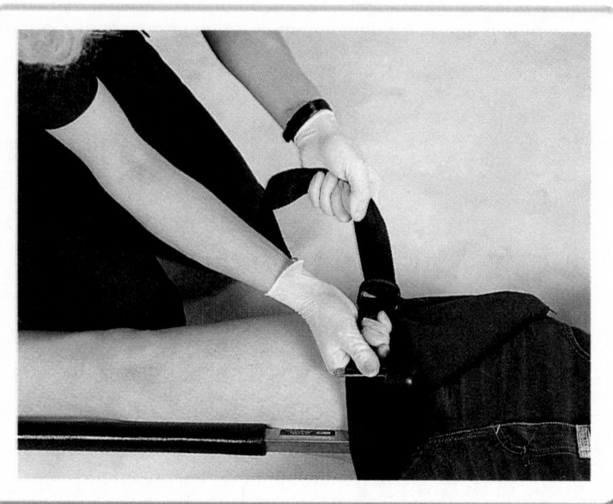

Step 6 Apply the ankle harness.

Position the ankle harness beneath the heel(s) and just above the ankle(s).

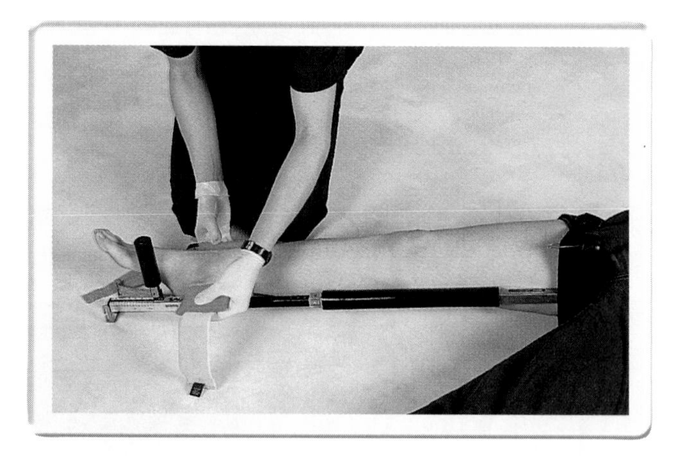

Fold down the number of comfort cushions needed to engage all of the ankle above the medial and lateral malleoli. Using the attached hook and loop straps, wrap the ankle harness around the ankle to secure snugly. Pull control tabs to engage the ankle harness tightly against the crossbar.

Step 7 Apply quantifiable dynamic traction.

Grasp the padded shaft of the splint with one hand and the traction handle with the other. Gently, extend the inner shaft of the splint until the desired amount of traction is recorded on the traction scale.

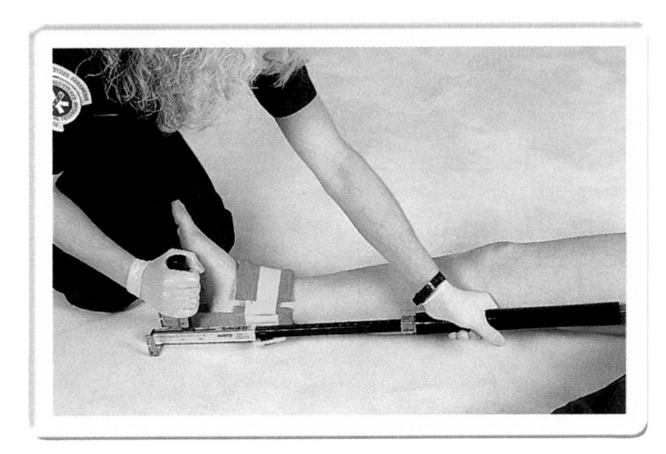

It is suggested to use 10% of the patient's body weight per fractured femur, up to 7 kg per leg (this would be a maximum of 14 kg for bilateral fractures).

Step 8 Position and secure the support straps. (Velcro straps may be applied directly over the fracture site.)

At the hollow of the knees, gently slide the large elastic leg cravat through and upward to the thigh. Repeat this process with the remaining cravats to minimize lower and midlimb movement.

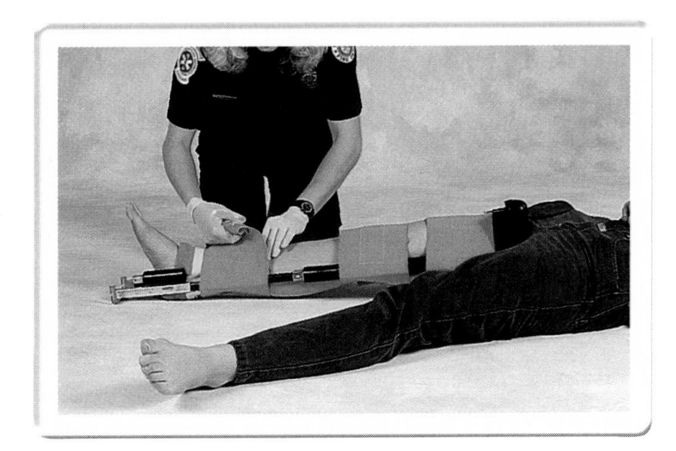

Step 9 Adjust straps and secure the feet.

Adjust the thigh strap at the upper thigh, making sure it is snug and secure. Ensure that all elastic leg cravats are secure. Apply the figure-of-eight strap around the feet to prevent rotation of the leg(s).

Step 10 Reevaluate the proximal and distal securing devices.

Inspect the splint to ensure the device is working properly. Check the ischial strap (proximal securing device) to ensure it remains secure. The ankle hitch (distal securing device) should be checked to ensure that it remains snug and pulls straight along the long axis of the leg and does not pull from the foot.

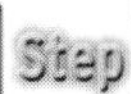 

Step 11 Reassess distal circulation, sensation, and motor function.

Check the distal circulation through pulse check or capillary refill. Also assess the distal sensation and motor function.

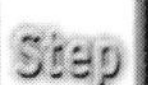 

Step 12 Secure the torso and splint to the long backboard to immobilize the hip.

Once the traction splint has been applied, the patient should be moved onto a long backboard and secured so that the end of the traction splint is supported.

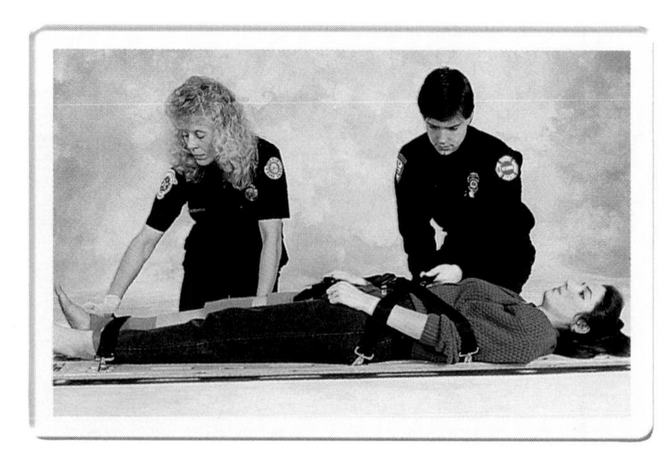

Pneumatic Antishock Garment

SKILL 37

Performance Objective

Given a patient, the candidate shall demonstrate the proper application of the pneumatic antishock garment (PASG) using the criteria herein described, in 5 minutes or less.

Equipment

The following equipment is required to perform this skill:

- Appropriate body substance isolation/personal protective equipment
- Pneumatic antishock garment
- Long backboard
- Long backboard straps

Indications

- Hemorrhagic shock
- Pelvic fracture
- Lower-body air splint

Contraindications

- Pulmonary edema
- Patients too large or too small for the device
  - Do not place children in a single leg of an adult device.
  - Use devices specifically designed for pediatric patients only if they fit the child. Increased abdominal pressure can cause severe respiratory compromise.

Complications

- Pulmonary edema
- Accelerated blood loss

Procedures

Step 1 Ensure body substance isolation before beginning procedures.

Prior to beginning patient care, appropriate body substance isolation procedures should be employed.

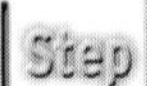 

Step 2 Assess distal circulation, sensation, and motor function.

Check the distal circulation through pulse check or capillary refill. Also assess the distal sensation and motor function.

In the Field

Use of Pneumatic Antishock Garments

There is little evidence to support the continued use of pneumatic antishock garments. If used at all, they should be used cautiously and limited to pelvic fractures or as a lower-body air splint.

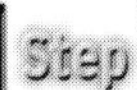 

Step 3 Prepare patient for application of the PASG.

Remove the patient's pants and underwear. Ensure that all sharp objects have been removed from the inflation area. This will prevent the garment from becoming damaged and unable to accomplish its purpose.

The PASG should be applied with the patient on a long backboard. Some systems will place the PASG on the board before the patient, then position it on the patient after the log roll.

Step 4 Position and align garment.

Place the PASG under the patient's legs and carefully work it under the patient's hip.

Remember: This patient may have a lower spine injury. Caution should be used to prevent additional injury.

Position the top of the garment below the patient's lower rib margin. The midsection should be carefully aligned along the spine.

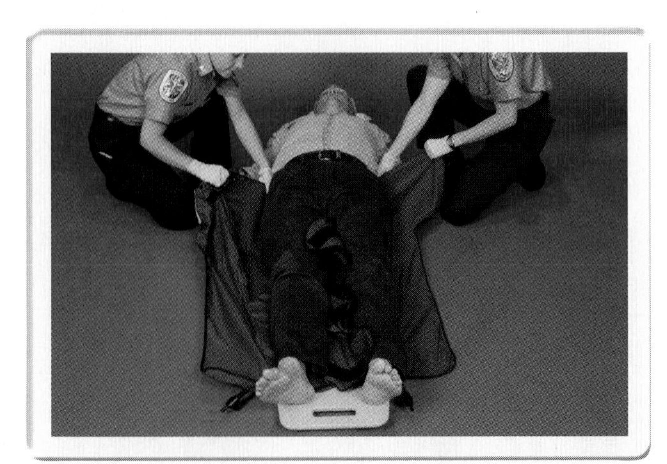

Step 5 Secure leg sections.

Snugly wrap the leg sections around the patient's legs. Because of the design of the garment, it will be easiest to wrap the left leg first and then the right. However, wrapping the right leg first will accomplish the same objective.

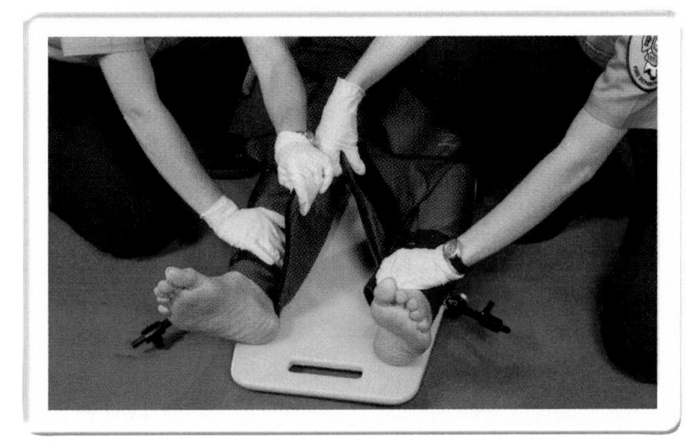

Step 6 Secure abdominal section.

Snugly apply the abdominal section. If the patient's abdomen is too large for the abdominal section to fit (from pregnancy, obesity, or just largeness), inflation of the garment is contraindicated in most situations.

Step 7 Inflate garment.

Close the valve to the abdominal section and ensure that the valves to both leg sections are open.

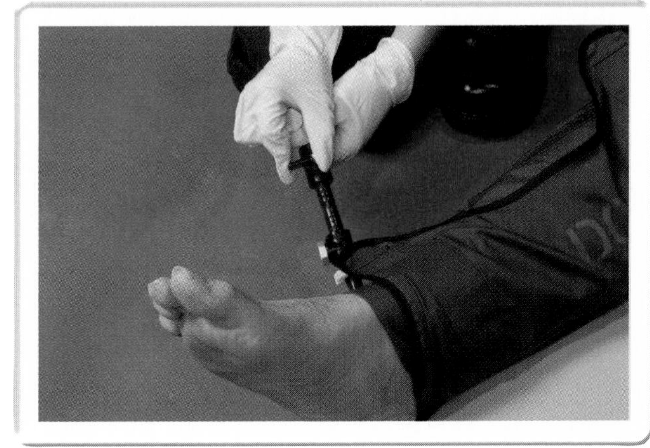

Using the foot pump, inflate both leg sections. Close the leg valves. Open the abdominal section valve and inflate. Close the abdominal valve. If desired, all three compartments can be inflated at one time. Stop inflating when the Velcro starts to crackle off the pop-off valve or the pop-off valve opens.

Step 8 Set valves to prevent loss of air from garment.

Recheck the valves for all sections of the PASG, making sure they are closed.

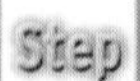

Step 9 Transport patient.

Use of a PASG on a patient requires transportation to a trauma center. Transport should begin without delay.

Check Your Knowledge

You are dispatched to a local nightclub for an assault victim. You arrive on scene to find a 22-year-old woman sitting on the curb in front of the club surrounded by a small group of people. Her left arm is wrapped in a t-shirt that is soaked with blood. After ensuring that the scene is safe and that you have taken the appropriate body substance isolation precautions, you approach the patient. She appears to be alert and oriented, and she tells you that another female attacked her with a knife, slashing her left arm. You remove the t-shirt to find an approximately 2" laceration to the patient's forearm that is bleeding profusely.

1. **If direct pressure to the laceration in the patient's arm and elevation of the extremity fail to control bleeding, what should be your next step?**
   - **A.** Apply a tourniquet proximal to the wound.
   - **B.** Apply an additional dressing to the wound without removing the first dressing.
   - **C.** Apply pressure to the brachial artery proximal to the wound.
   - **D.** Apply an inflatable splint.
2. **When bandaging this patient's injury it is important to remember to:**
   - **A.** check distal pulse and/or capillary refill.
   - **B.** use enough pressure to slow circulation.
   - **C.** make sure the wound is not covered completely with a dressing.
   - **D.** none of the above.
3. **The proper position for transporting this patient would be:**
   - **A.** Trendelenberg position.
   - **B.** recovery position.
   - **C.** semi-Fowler's position.
   - **D.** immobilization on a long backboard.
4. **When should you check for a distal pulse?**
   - **A.** Before bandaging the extremity
   - **B.** After bandaging the extremity
   - **C.** Any time that is convenient
   - **D.** Both A and B

Additional Questions

5. **When bandaging an eye injury, why should you also cover the uninjured eye?**
   - **A.** To keep the patient from looking around
   - **B.** To keep blood and fluids out of the uninjured eye
   - **C.** To keep the patient from getting a headache
   - **D.** To prevent sympathetic movement
6. **For which of the following wounds would you use an occlusive dressing?**
   - **A.** Neck laceration
   - **B.** Scalp laceration
   - **C.** Joint laceration
   - **D.** Compound fracture
7. **Under which of the following conditions should you remove an impaled object?**
   - **A.** The object interferes with adequate CPR.
   - **B.** The object is dirty.
   - **C.** The object causes the patient severe pain.
   - **D.** You never remove an impaled object.
8. **When transporting an amputated body part, you should:**
   - **A.** keep the body part warm and dry.
   - **B.** pack the body part in ice.
   - **C.** soak the body part in sterile saline.
   - **D.** none of the above.
9. **The proper technique for splinting a joint injury includes:**
   - **A.** securing and stabilizing the bones above and below the injury site.
   - **B.** securing and stabilizing the entire injured extremity.
   - **C.** reassessing distal circulation, sensation, and motor function after splinting.
   - **D.** all of the above.
10. **In what position should you splint an elbow injury?**
    - **A.** Bend the arm against the body and secure with a sling and swathe.
    - **B.** Straighten the arm and secure to a board splint.
    - **C.** Splint the injury in the position you find it.
    - **D.** You should not splint an elbow injury.

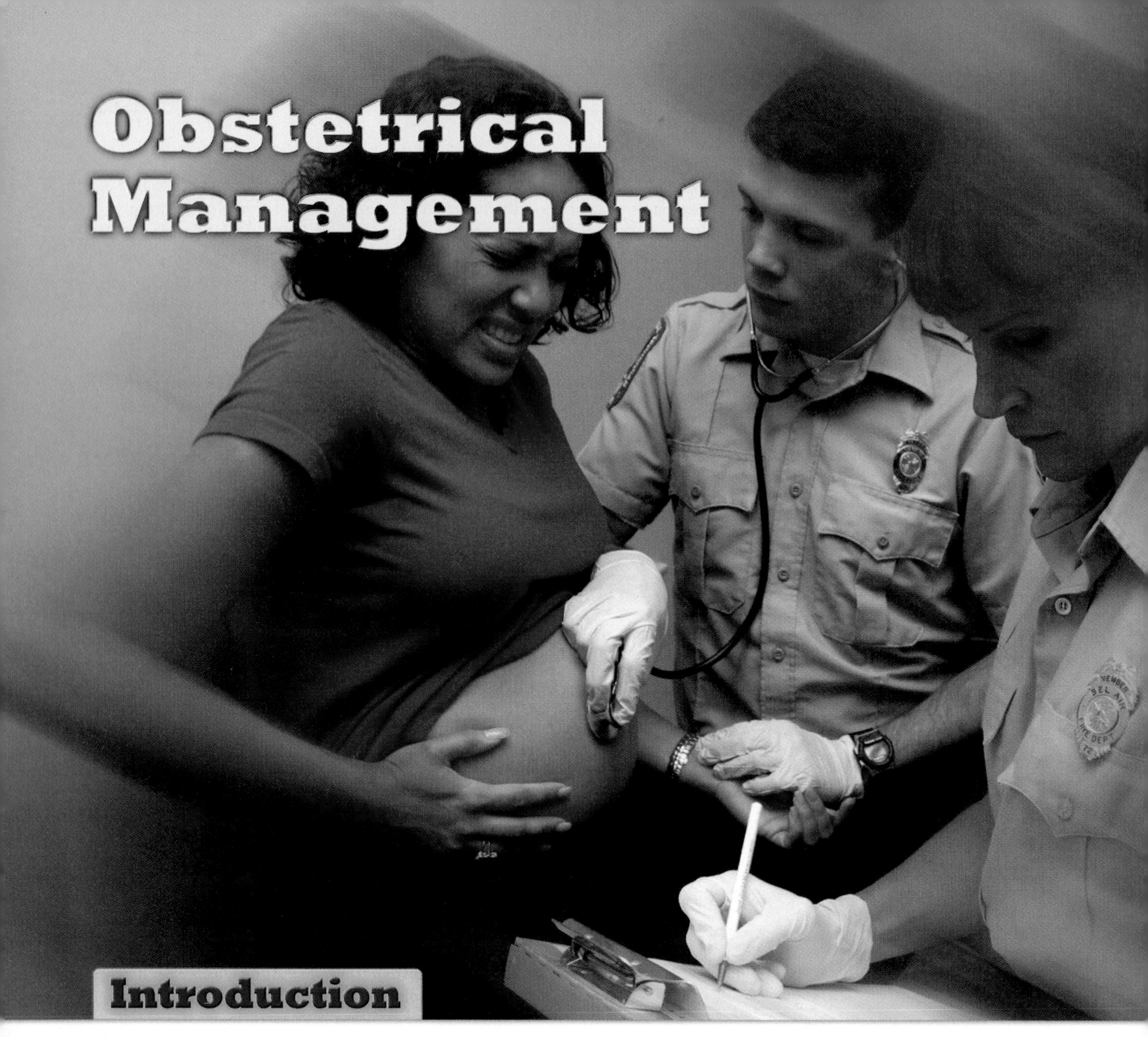

Obstetrical Management

Introduction

It is important to maintain the knowledge and skills to manage childbirth, potential complications of delivery, and the postdelivery care of the newborn. Although childbirth is a natural process that will occur with or without you, a patient could require your assistance at any given time.

The skills in this section demonstrate the proper assessment, preparation, and procedures for natural childbirth as well as the management of a breech delivery and a prolapsed cord.

SECTION 6

SKILL 38

Childbirth

Performance Objective

Given a pregnant patient with a term pregnancy and all necessary equipment, the candidate shall demonstrate the proper assessment, preparation, and procedures for the delivery of an infant, within 10 minutes or less.

Equipment

The following equipment is required to perform this skill:

- Appropriate body substance isolation/personal protective equipment
- Obstetrical kit, to include:
  - Bulb syringe
  - Umbilical clamps or ties (three)
  - Scalpel or umbilical shears
  - Sterile gloves
  - Drapes
  - Abdominal pads
  - 5″ x 9″ absorbent pads (chucks)
  - Foil blanket
  - 4″ x 4″ gauze pads
  - Plastic bag (for placenta), with tie

Equipment that may be helpful:

- Clean sheets
- Multi-trauma dressings
- Receiving blanket
- Infant knit cap

Indications

- Natural childbirth when delivery is imminent
  - Delivery is imminent when the baby's head is visible in the birth canal (also known as crowning).

Contraindications

- Complications of delivery
- Single footling breech
- Limb presentation
- Prolapsed cord

Complications

Maternal

- Perineal tear
- Uterine rupture
- Postpartum hemorrhage

Neonatal

- Breech birth
- Nuchal cord
- Prolapsed cord

Procedures

Step 1 Ensure body substance isolation before beginning procedures.

Prior to beginning patient care, appropriate body substance isolation procedures should be employed.

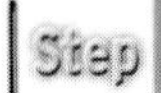

Step 2 Prepare equipment and position mother.

Open the obstetrical kit and check to make sure all the necessary equipment is present. This kit should contain draping for the mother, a scalpel, at least three umbilical cord ties or clamps, sterile gloves, a foil blanket, a bulb syringe, a plastic bag for the placenta, several 4″ x 4″ gauze pads, and some 5″ x 9″ absorbent pads (chucks).

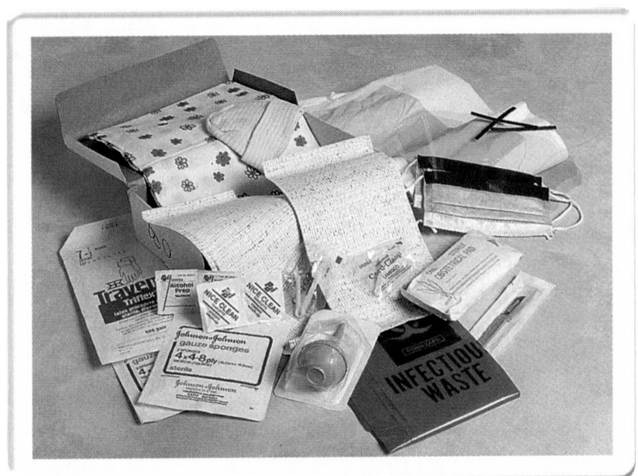

Prepare for the delivery by opening a 4″ x 4″ gauze pad and unwrapping it to 8″ x 8″. Push the tip of the bulb syringe through the gauze pad and tie the four corners around the top. This will give you a better grip on the wet rubber syringe when it is time to use it.

Position the mother near the end of a bed if possible. Have her lie supine with her knees bent and feet spread apart. Administer oxygen by nasal cannula or nonrebreathing mask. If time permits, an IV should be started with normal saline or lactated Ringer's.

In the Field

Childbirth and Emergency Childbirth

Childbirth is a natural event that is not by nature an emergency. For millions of years, mammals, including humans, have been having babies with little need for medical attention.

In some cases, however, childbirth can pose a serious danger to the mother, the baby, or both. A detailed history should reveal known complications. In these cases, emergency transport should be performed in order to reduce the risks of field delivery. Attempting to physically delay delivery can be very dangerous and should never be attempted without a physician's order.

Step 3 Position barrier devices and draping.

All personnel who will be in the immediate area of the delivery should follow full body substance isolation precautions. Gloves, goggles, and masks should be donned as soon as possible and before any contact with the perineum or baby.

If time permits, place the drapes over the mother's legs and abdomen. The chuck should be placed under the buttocks.

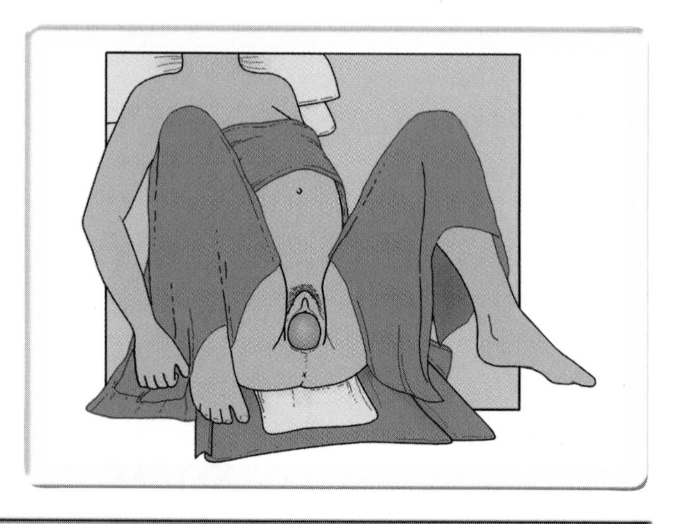

Step 4 Coach mother and communicate with team members.

Until the baby's head is visible in the vaginal opening, transport should be continued. No effort to get the mother to push should be made.

As the head begins to appear in the vaginal opening, inform the mother that the baby is coming. Have her push with the contractions, and stop and breathe deeply after each push. It is a good idea to utilize the training of the mother's delivery coach if that coach is present.

Open communications should be given so that the mother and all team members know the progress of the delivery and can anticipate what steps will need to be performed next.

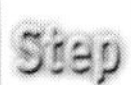

Step 5 Guide head and prevent explosive delivery.

As the head delivers, apply gentle pressure to the perineum and presenting part to prevent an explosive delivery.

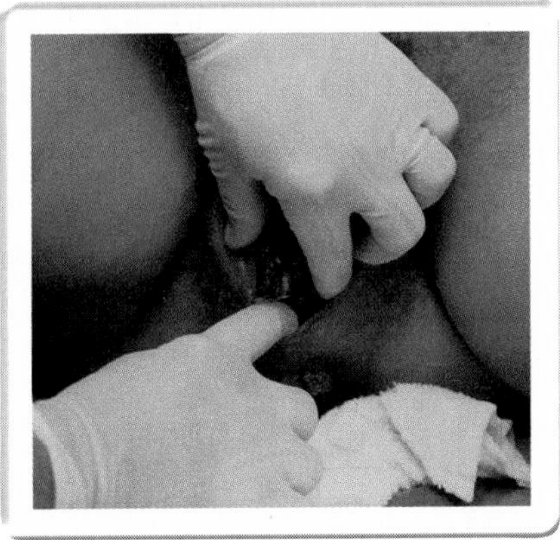

If the amniotic sac has not ruptured it may be necessary to break the membrane with your finger. Keep the baby's head up to prevent contact with any vaginal discharge, urine, or fecal material that may be present.

If the umbilical cord is around the baby's neck, attempt to reposition it by lifting it over the baby's head. If you are unable to do this, apply two umbilical cord clamps and cut the cord. Continue with normal delivery.

Step 6 Suction baby's mouth and nose.

Once the baby's head is completely delivered, have the mother stop pushing and rest for a short time. Have her breathe easily, or pant if necessary. Suction the baby's mouth and nose with a bulb syringe and check for the presence of meconium. Since babies are obligate nose breathers, the mouth should always be cleaned first to prevent aspiration of mucus in the oropharynx.

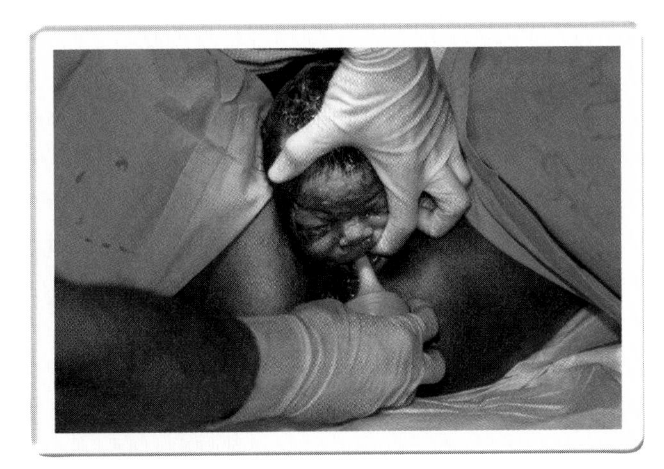

Step 7 Continue with delivery.

Continue with the delivery by having the mother continue to push with each contraction. This final step should be very quick. You will need to pay close attention to the baby and maintain a firm grasp. Record the time of birth.

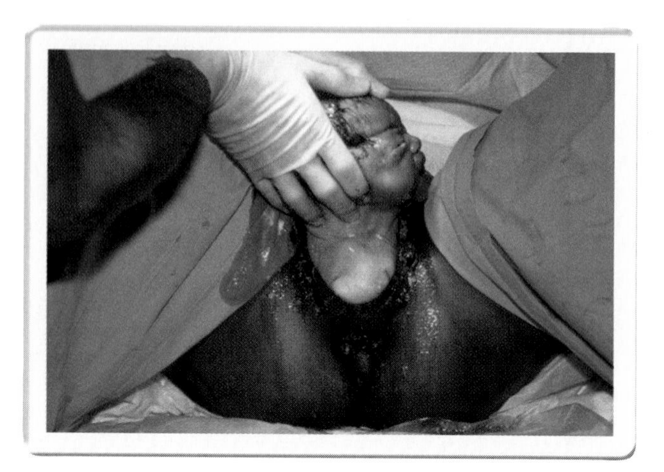

Step 8 Suction baby's mouth and nose.

Lay the baby down and quickly suction the mouth and nose again.

Step 9 Cut and clamp cord.

Apply two umbilical cord clamps or ties approximately 8″ to 10″ from the baby, 2″ to 3″ apart. Cut between the clamps. A slight amount of bleeding is expected initially. However, if the bleeding continues, especially from the baby's end, a second cord clamp or tie should be applied near the first. It is a good habit to apply a tie below the clamp on the baby's cord as an added safety measure.

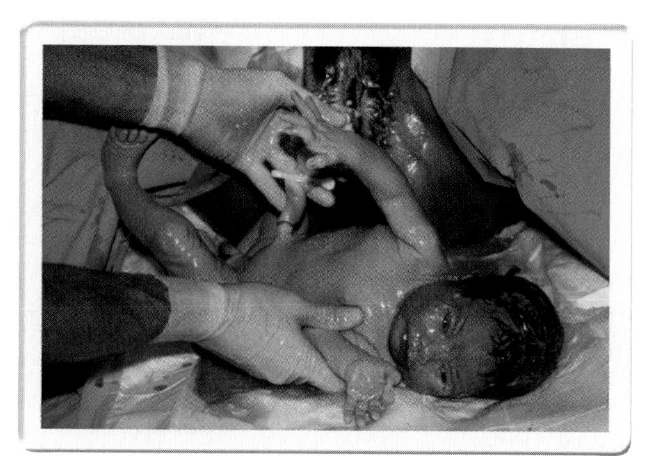

Step 10 Dry the baby.

Dry the baby of amniotic fluid without removing the vernix and wrap him or her in a blanket. The blanket should be positioned to cover the top of the baby's head to prevent heat loss. Placing the baby on her or his side with the head slightly lower than the rest of the body will facilitate drainage and ease breathing.

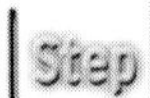

Step 11 Stimulate breathing.

In most cases, the delivery and drying process will be enough to stimulate breathing. If breathing does not start soon after birth, stimulate respiratory effort by rubbing the baby's back or by flicking or tapping the soles of the feet. Further, more aggressive stimulation is not necessary. If this action does not stimulate breathing, more advanced supportive care is required (see the box entitled *Neonatal Resuscitation* on the following page).

Step 12 Evaluate baby and mother, and deliver placenta.

The baby should be evaluated using the APGAR score (see the table on the next page). A measurement should be taken after 1 minute and again 5 minutes later.

Once the baby is delivered and cared for, evaluate the mother. Evaluate the perineum for tears and vaginal bleeding. Perineal tears should be dressed while waiting for placental delivery. Bleeding from the vaginal opening could be coming from the vaginal wall or from the uterus. Uterine bleeding can be controlled by fundal massage or having the baby nurse.

The placenta should deliver in 5 to 15 minutes, but transport should not be delayed waiting for it.

Keep the placenta as clean as possible and place it in a plastic bag. The placenta should be delivered to the hospital with the mother and baby.

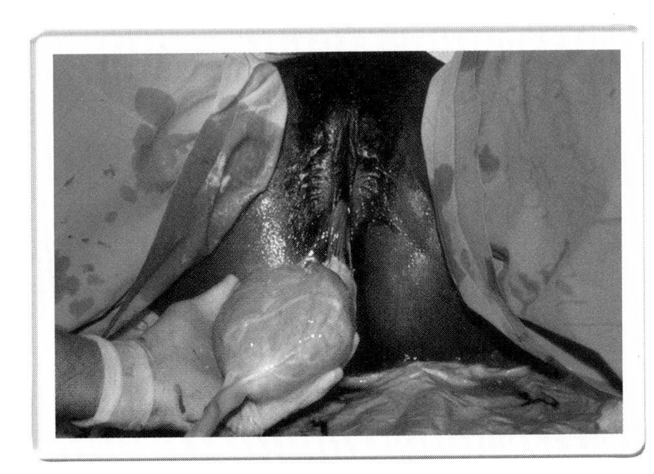

The APGAR Score

Condition	Description	Score
Appearance–skin color	Completely pink	2
	Body pink, extremities blue	1
	Centrally blue, pale	0
Pulse rate	Over 100	2
	Less than 100	1
	Absent	0
Grimace–irritability	Cries	2
	Grimaces	1
	No response	0
Activity–muscle tone	Active motion	2
	Some flexion of extremities	1
	Limp	0
Respiratory–effort	Strong cry	2
	Slow and irregular	1
	Absent	0

In the Field

Neonatal Resuscitation

After delivery of a baby, it is important to perform an adequate assessment to determine if he or she is doing well. A simple APGAR usually is not sufficient to determine the need for resuscitation. The American Heart Association and the American Academy of Pediatrics have come together to create the neonatal resuscitation pyramid. This pyramid guides the process of resuscitation of a struggling newborn. Although not all newborns will need the full gamut of resuscitation efforts, all newborns need the procedures specified in the top level of the pyramid.

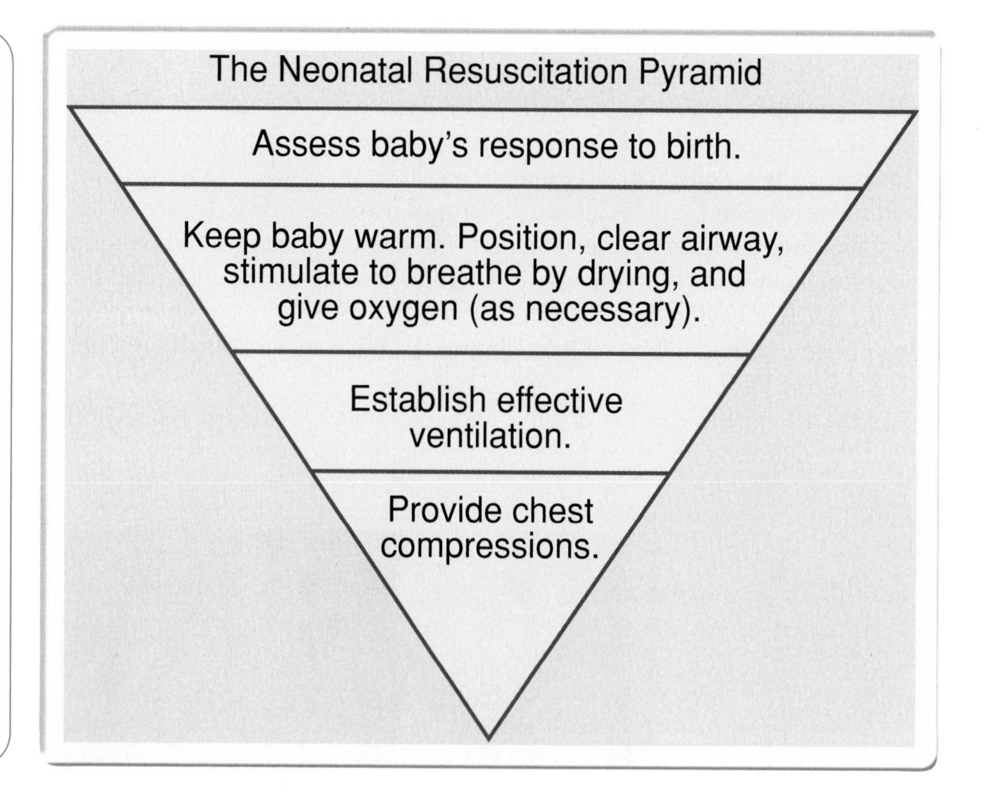

Step 13 Properly dispose of contaminated equipment.

Dispose of any sharp instruments with blood or body fluid contamination in puncture-resistant sharps containers. Other material that has been contaminated by blood or body fluids should be disposed of in biohazard bags. Waste materials that are not contaminated by blood or body fluids can be disposed of in normal trash receptacles.

Step 14 Document.

Document the delivery of the baby on the mother's patient report. However, the baby should have a separate patient report that represents the assessment and any procedures performed during the care of the newborn.

SKILL 39

Breech Delivery

Performance Objective

Given a pregnant patient with breech presentation, the candidate shall demonstrate the assessment and procedures for a breech delivery, within 10 minutes or less.

Equipment

The following equipment is required to perform this skill:

- Appropriate body substance isolation/personal protective equipment
- Obstetrical kit, to include
  - Bulb syringe
  - Umbilical clamps or ties (three)
  - Scalpel or umbilical shears
  - Sterile gloves
  - Drapes
  - Abdominal pads
  - 5″ x 9″ absorbent pads (chucks)
  - Foil blanket
  - 4″ x 4″ gauze pads
  - Plastic bag (for placenta), with tie

Equipment that may be helpful:

- Clean sheets
- Multi-trauma dressings
- Receiving blanket
- Infant knit cap

Indications

- Breech presentation that fails to continue delivery once the legs and body have delivered

Contraindications

- None in emergency situations

Complications

- None in emergency situations

Procedures

 1 **Ensure body substance isolation before beginning procedures.**

Prior to beginning patient care, appropriate body substance isolation procedures should be employed.

  **Prepare equipment and position mother.**

Open the obstetrical kit and check to make sure all the necessary equipment is present. This kit should contain draping for the mother, a scalpel, at least three umbilical cord ties or clamps, sterile gloves, a foil blanket, a bulb syringe, a plastic bag for the placenta, several 4″ x 4″ gauze pads, and some 5″ x 9″ absorbent pads (chucks).

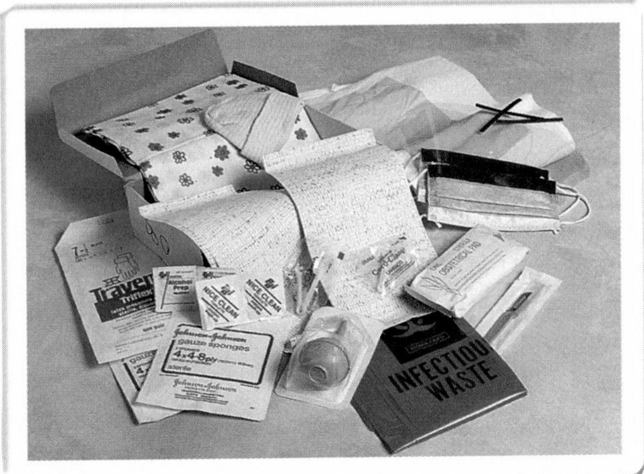

Prepare for the delivery by opening a 4″ x 4″ gauze pad and unwrapping it to 8″ x 8″. Push the tip of the bulb syringe through the gauze pad and tie the four corners around the top. This will give you a better grip on the wet rubber syringe when it is time to use it.

Position the mother near the end of a bed if possible. Have her lie supine with her knees bent and feet spread apart. Administer oxygen by nasal cannula or nonrebreathing mask.

Position barrier devices and draping.

Field delivery of a breech birth is not desirable because the possibility of complications is extremely high. Cesarean delivery is preferred if delivery can be delayed until arrival at the hospital.

All personnel who will be in the immediate area of the delivery should follow full body substance isolation precautions. Gloves, goggles, and masks should be donned as soon as possible and before any contact with the perineum or baby.

If time permits, place the drapes over the mother's legs and abdomen. The chuck should be placed under the buttocks.

Coach mother and communicate with team members.

Until the presenting part is visible in the vaginal opening, transport should be continued. Make every effort to discourage the mother to push by having her pant or blow through the contractions.

As the feet or buttocks appear in the vaginal opening, inform the mother that the baby is coming, and that the presentation is breech.

Do not encourage pushing until the option of delaying delivery no longer exists. At that point, have the mother push with the contractions, and stop and breathe deeply after each push. It is a good idea to utilize the training of the mother's delivery coach if that coach is present.

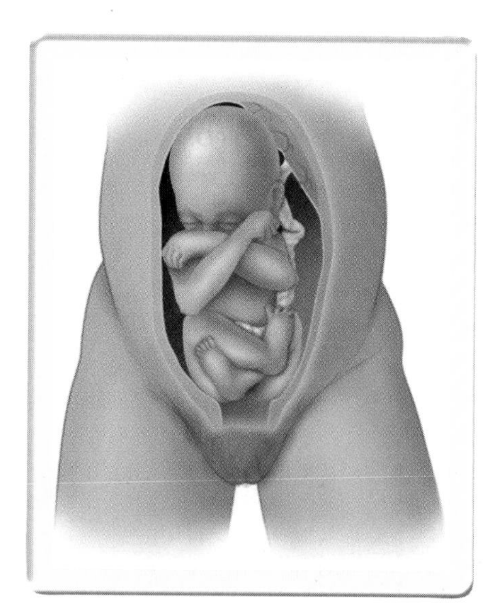

Open communications should be given so that the mother and all team members know the progress of the delivery and can anticipate what steps will need to be performed next.

Step 5 Guide baby and prevent explosive delivery.

As the feet and body deliver, gently control the mother's pushing force with gentle pressure to the perineum and presenting part to prevent an explosive delivery.

If the amniotic sac has not ruptured it may be necessary to break the membrane with your finger. Keep the baby up to prevent contact with vaginal discharge, urine, or fecal material that may be present.

If the umbilical cord is around the baby's neck, attempt to reposition it by lifting it over the body. If you are unable to do this, evaluate the blood flow through the cord and the tightness around the neck. Cutting the cord may remove the most reliable oxygen supply the baby has.

Step 6 Consider need for immediate transport.

Since the head will require more time to deliver than the body, consider whether immediate transport should begin. If the head does not deliver in the next 3 to 5 minutes, transport will be the only acceptable option. Continue to Step 8.

Step 7 Gently guide head.

If the head delivers, continue with care of the baby and mother as in a normal cephalic delivery (see Skill 38).

Step 8 Create airway.

If after 3 to 5 minutes the head has not delivered, an airway into the vagina must be made. With a fresh sterile glove, place two fingers, palm toward the baby, into the mother's vagina to the level of the baby's nose. Spread the fingers apart to form a V. If possible, pull the vaginal wall away from the baby's face. You may have to hold this position throughout transport and into the delivery room.

Step 9 Assess further needs.

Assess further needs for the baby and the mother. In most cases, rapid transport and notification of the breech presentation to the receiving facility are all that is required.

SKILL 40

Prolapsed Cord

Performance Objective

Given a pregnant patient with cord presentation, the candidate shall demonstrate the assessment and procedures for a prolapsed cord, within 10 minutes or less.

Equipment

The following equipment is required to perform this skill:

- Appropriate body substance isolation/personal protective equipment
- Obstetrical kit, to include
  - Bulb syringe
  - Umbilical clamps or ties (three)
  - Scalpel or umbilical shears
  - Sterile gloves
  - Drapes
  - Abdominal pads
  - 5" x 9" absorbent pads (chucks)
  - Foil blanket
  - 4" x 4" gauze pads
  - Plastic bag (for placenta), with tie

Equipment that may be helpful:

- Clean sheets
- Multi-trauma dressings
- Receiving blanket
- Infant knit cap

Indications

- Presentation of the umbilical cord prior to the delivery of the baby

Contraindications

- None in emergency situations

Complications

- None in emergency situations

Procedures

Step 1 Ensure body substance isolation before beginning procedures.

Prior to beginning patient care, appropriate body substance isolation procedures should be employed. All personnel who will be in the immediate area of the delivery should follow full body substance isolation precautions. Gloves, goggles, and masks should be donned as soon as possible and before any contact with the perineum or umbilical cord.

Step 2 Assess presenting part.

Assess the presenting part to determine the amount of umbilical cord present. Gently assess the cord and determine the presence of a pulse. Prepare for immediate transport.

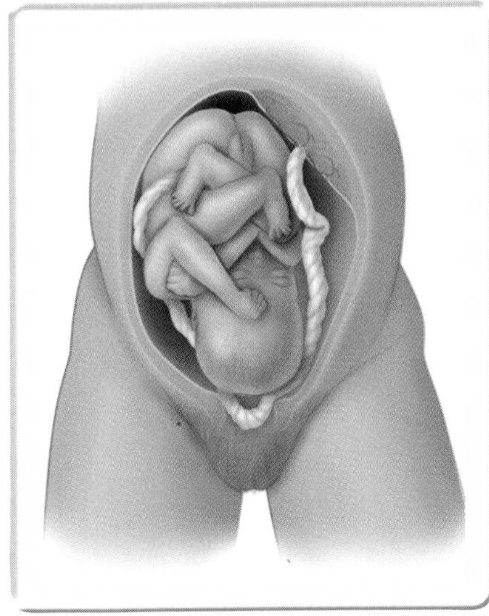

Step 3 Position mother and apply oxygen.

Position the mother in a high Trendelenburg position (supine with lower extremities elevated approximately 12″) or a knee-chest position (see note), and coach the mother not to push. Elevation of the inferior uterus is essential to shift the weight of an engaging fetus off the cervix. Administer 100% supplemental oxygen via nonrebreating mask.

Note: Use of the knee-chest position in the back of a moving ambulance can be extremely hazardous. The high Trendelenburg position is preferred while in transport.

With a prolapsed cord, the mother must be in the knee-chest position in order to perform Step 4. It will be important for the vehicle operator to understand the precarious position of the patient.

Step 4 Place gloved hand into vaginal opening and push baby's head back into uterus.

With a sterile glove, place two fingers, palm toward baby, into the vaginal opening. Push the baby's head back into the uterus and remove pressure from the cord. Reassess the umbilical pulse. Cover any exposed cord with moistened dressings. It will be necessary to hold this position throughout transport.

Check Your Knowledge

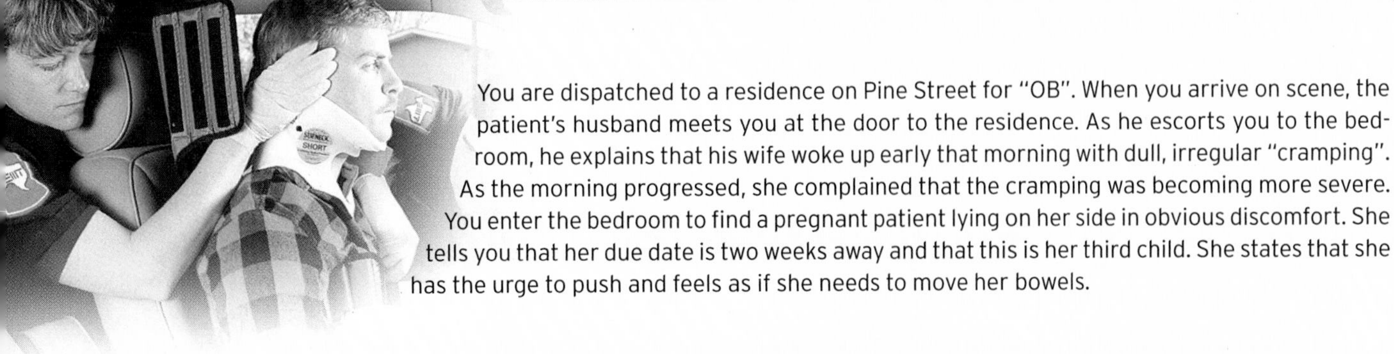

You are dispatched to a residence on Pine Street for "OB". When you arrive on scene, the patient's husband meets you at the door to the residence. As he escorts you to the bedroom, he explains that his wife woke up early that morning with dull, irregular "cramping". As the morning progressed, she complained that the cramping was becoming more severe. You enter the bedroom to find a pregnant patient lying on her side in obvious discomfort. She tells you that her due date is two weeks away and that this is her third child. She states that she has the urge to push and feels as if she needs to move her bowels.

1. **What should your first step be when preparing to examine this patient?**
   A. Place the mother in the proper position, lying supine near the edge of the bed with her knees bent and her feet spread apart.
   B. Open a sterile OB kit and prepare your equipment.
   C. Place drapes over the mother's legs and abdomen and place a chuck under her buttocks.
   D. Don the proper body substance isolation equipment, including gloves, goggles, and mask.

2. **When would it be appropriate to delay transport and proceed with delivery on scene?**
   A. When the mother feels the urge to push
   B. When the baby's head appears at the vaginal opening
   C. When the amniotic sac breaks
   D. When the mother expels the mucous plug

3. **As the infant's head begins to deliver, why is it important to apply gentle pressure to the perineum and presenting part?**
   A. To avoid an explosive delivery
   B. To delay delivery until you get to the hospital
   C. To make the mother more comfortable
   D. To prevent the mother from having a bowel movement

4. **Why is it important to suction the infant's mouth before the nose?**
   A. So the infant can cry more easily
   B. To avoid aspiration of mucous into the oropharynx
   C. Because it is more efficient
   D. All of the above

5. **Under what circumstances should you consider clamping and cutting the umbilical cord before delivery is complete?**
   A. If the baby is breech
   B. If the baby is not breathing
   C. If the amniotic sac has not ruptured
   D. If the cord is wrapped around the baby's neck and you cannot reposition it

Additional Questions

6. **When preparing to cut the umbilical cord, where should you position the clamps?**
   A. 4" to 6" from the baby
   B. 6" to 8" from the mother
   C. 8" to 10" from the baby
   D. 10" to 12" from the mother

7. **In addition to pulse rate and respiratory effort, the Apgar score evaluates:**
   A. appearance.
   B. body temperature.
   C. pupil size.
   D. all of the above.

8. **If the infant's head does not deliver within 3 to 5 minutes in case of a breech delivery, it may be necessary to:**
   A. gently pull on the baby to facilitate delivery.
   B. gently push on the mother's abdomen to facilitate delivery.
   C. create an airway into the vagina.
   D. all of the above.

9. **What is the proper position for transporting the mother in the case of a prolapsed cord?**
   A. Fowler's position
   B. Recovery position
   C. Supine
   D. Trendelenberg or knee-chest position

10. **How can you remove pressure from a prolapsed cord during transport?**
   A. Use two fingers to push and hold the baby's head back into the uterus.
   B. Gently pull the cord out of the vagina.
   C. Clamp the cord.
   D. None of the above.

Operational Skills

Introduction

Operational skills are performed on almost every call and many of the skills benefit the responder more than the patient. The number one responsibility of the responder is personal safety. In fact, in order of importance, the responder comes first, bystanders and family second, and the patient third. If a responder gets injured, it takes away from the number of responders available to care for the patient; if bystanders or the patient's family get injured, it takes additional responders away from the patient.

The patient's best interests are served when responders take care of themselves and family first. The skills in this section demonstrate how to perform operational skills while keeping responders, bystanders, and patients safe.

SECTION 7

Cot Operation

Performance Objective

Given a cot and a patient, the candidates will be able to secure the patient comfortably to the cot, raise the cot to a transport position, move the cot to the ambulance, raise the cot to the loading position, and load and unload the cot, in 10 minutes or less.

Equipment

The following equipment is required to perform this skill:

- Appropriate body substance isolation/personal protective equipment
- Multi-level ambulance cot

Indications

- Transport of a patient on a cot

Contraindications

- Patients above the maximum weight rating of the cot

Complications

- Back injury to the operator

Procedures

Ensure body substance isolation before beginning procedures.

Prior to beginning patient care, appropriate body substance isolation procedures should be employed.

Secure patient.

With the patient positioned on the cot, gently raise the head of the cot to a position of comfort not to exceed 45°. Place straps on the patient's chest and thighs and secure.

Raise cot for transport.

Lifting from the head and foot of the patient, raise the cot to position three, four, or five.

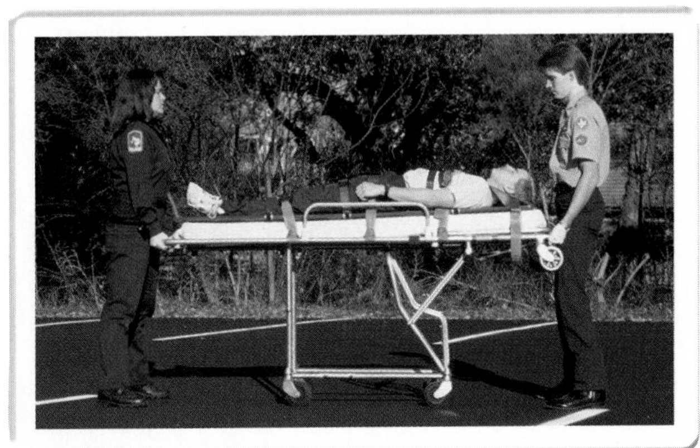

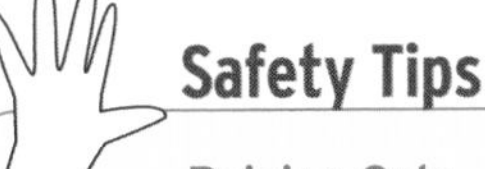

Safety Tips

Raising Cots

Raising the cot to a higher position is not recommended for routine transport of patients. These higher positions are not stable should sudden shifting of the patient occur.

Move the patient on the cot.

With one person at the foot and another at the head, move the patient with the cot to the desired location. It is usually best to move the patient feet first, but on inclines and stairs always keep the patient's head up and feet down.

Step 5 Raise cot to loading position and load cot.

Lifting from the head and foot of the patient, raise the cot to the appropriate loading position.

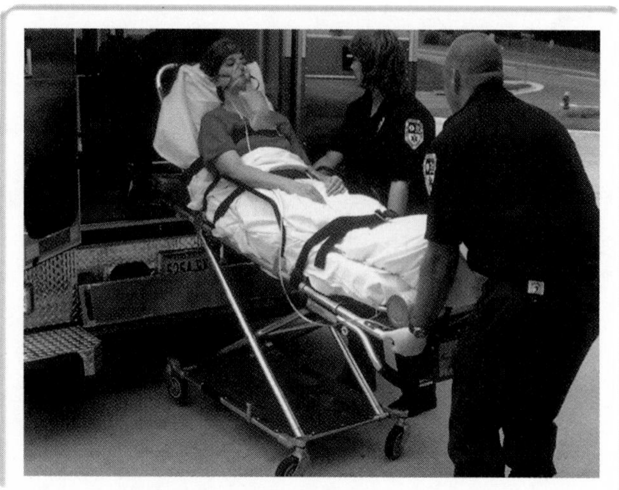

Place the back wheels onto the rear platform of the ambulance. While the person at the patient's foot lifts the cot and releases the locking mechanism, the second person lifts the undercarriage.

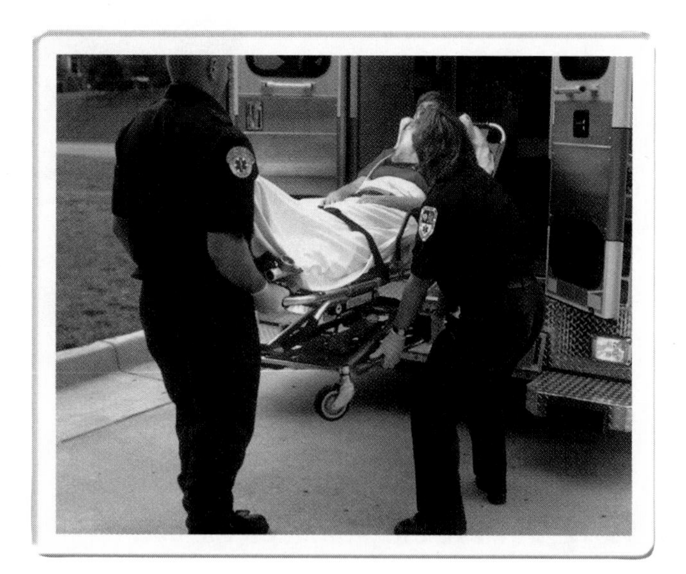

continued

Roll cot into the ambulance and lock it in place.

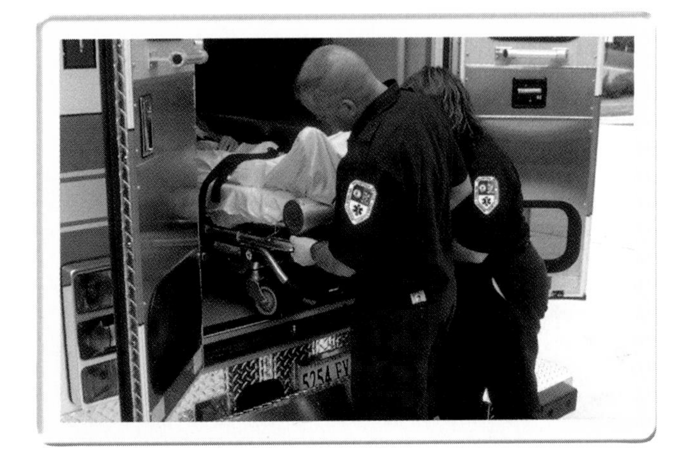

Unload cot.

Release the locking mechanism for the cot. With one person positioned at the patient's feet, begin unloading the cot by lifting and rolling the cot. The second person should guide the undercarriage to the ground and ensure the wheels have locked before completely removing the cot from the ambulance.

Patient Lifting Techniques

Performance Objective

Given a patient in a variety of positions on the ground, chair, or bed, the candidates shall demonstrate safe lifting techniques, in 1 minute or less.

Equipment

The following equipment is required to perform this skill:

- Appropriate body substance isolation/personal protective equipment
- Multi-level ambulance cot
- Standard bed sheet

Indications

- Movement of a patient

Contraindications

- None

Complications

- Improper lifting can result in injury to rescuers or the patient.

Procedures

Standard Lift

This technique is used to lift a cot, backboard, scoop stretcher, or other heavy loads.

Step 1 Ensure body substance isolation before beginning procedures.

Prior to beginning patient care, appropriate body substance isolation procedures should be employed.

Step 2 Squat using proper body mechanics.

With your feet spread comfortably, about shoulder width apart, squat into position by bending the knees. Keep your back straight.

Step 3 Grasp the object to be lifted.

Grasp the cot or other object to be lifted with your palms facing out. Keep your arms close to the center of your body.

Step 4 Perform lift.

Lift the cot or other object by straightening the knees. The back should remain straight, and your arms should remain close to your body.

Step 5 Lower patient.

To lower the patient, reverse the steps.

Two-Person Extremity Lift Technique

This technique is used to lift a non-trauma patient from the ground to the cot.

Ensure body substance isolation before beginning procedures.

Prior to beginning patient care, appropriate body substance isolation procedures should be employed.

First rescuer: Raise patient's upper body.

Kneeling behind the patient, grasp the patient's wrists. Hold the arms into the chest. Pull the patient into a seated position.

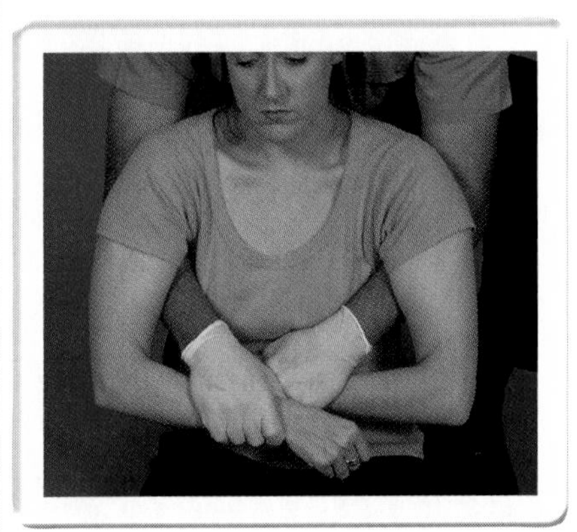

Second rescuer: Prepare to lift the lower body.

Facing the same direction as the patient, place your hands under the patient's knees.

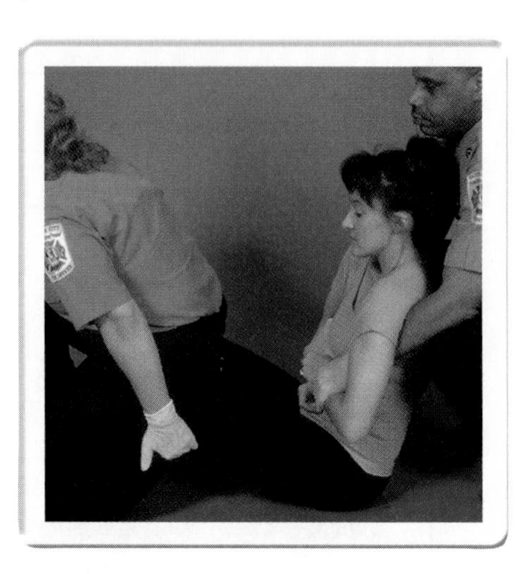

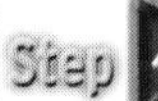

Both Rescuers: Lift the patient.

Working together and using the standard lifting technique, lift the patient.

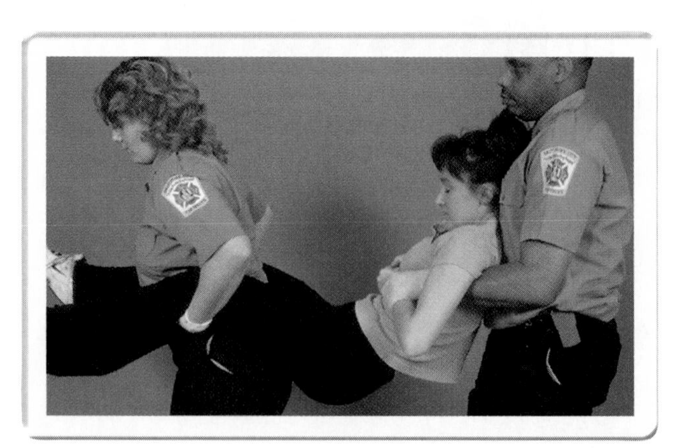

Position the patient.

Carefully lower the patient onto the cot. Position as appropriate and secure.

Bed-to-Cot Transfer: Sheet Technique

This technique is used to move a non-trauma patient from a bed to the cot using the sheet technique. The procedure requires a minimum of two people. With larger patients, as many as six people may be required.

Ensure body substance isolation before beginning procedures.

Prior to beginning patient care, appropriate body substance isolation procedures should be employed.

Prepare the sheet.

Using a standard bed sheet or a draw sheet, roll the edges to make a handle.

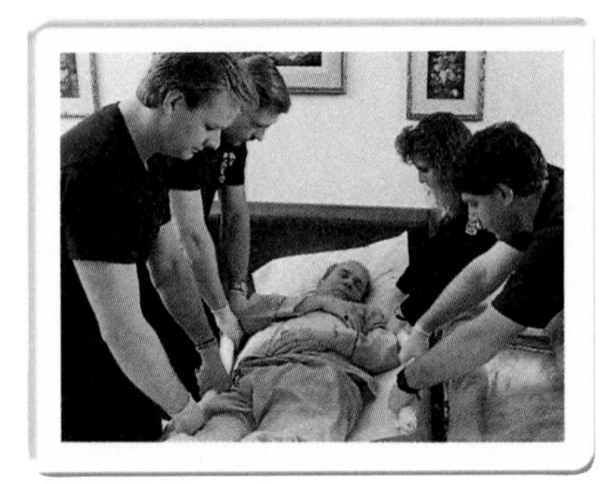

Position patient.

Lifting as a team, move the patient to the edge of the bed. Ensure the patient does not roll out of the bed before the transfer is complete.

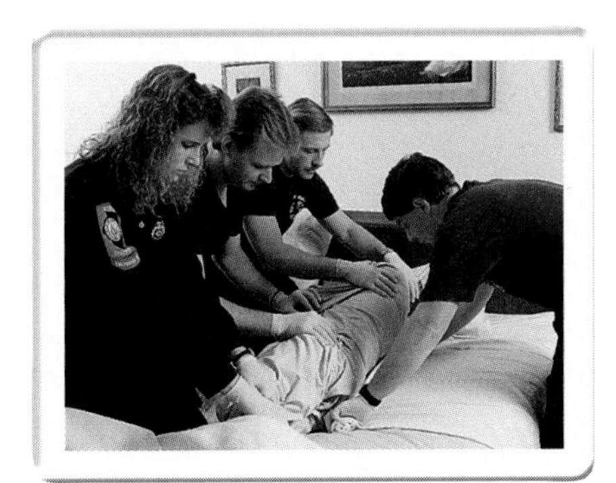

Step 4 Move and position the patient.

As before and working as a team, lift the patient carefully onto the cot. Position patient as appropriate and secure.

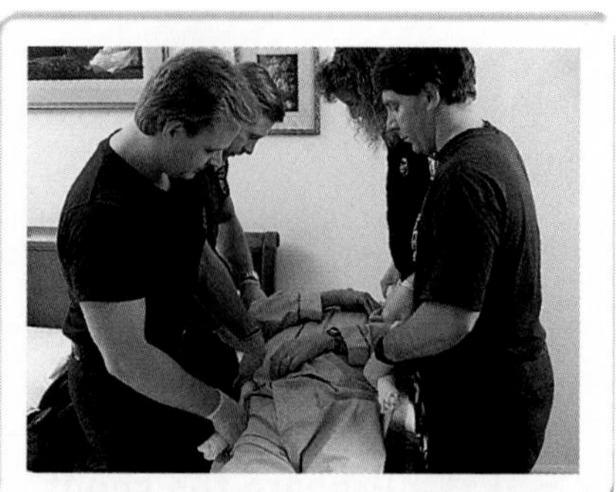

Bed-to-Cot Transfer: Direct Lift Technique

This procedure is used to move a non-trauma patient from a bed to the cot using the direct lift technique. This requires a minimum of two people. With larger patients, three people may be required.

Step 1 Ensure body substance isolation before beginning procedures.

Prior to beginning patient care, appropriate body substance isolation procedures should be employed.

Step 2 Position the cot.

Set up for this transfer by placing the head of the cot at the patient's feet, or the foot of the cot at the patient's head. This will allow you to lift the patient, rotate 90°, and place the patient.

Step 3 Prepare patient for move.

Working as a team, place your arms under the patient. Be sure to support the head and shoulders, the midback, the hips, and the lower legs.

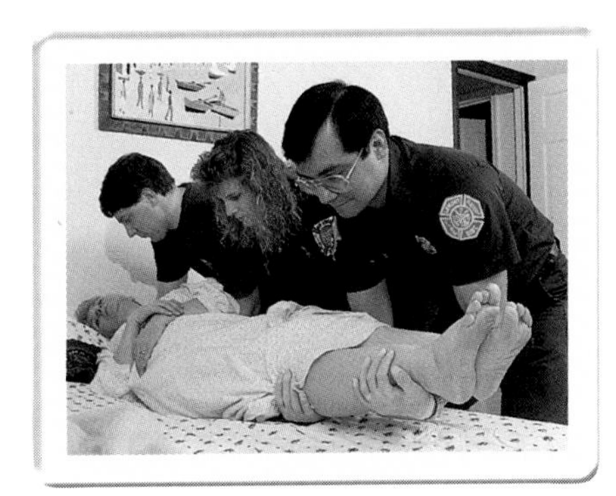

Step 4 Lift patient.

Working as a team, lift the patient by rolling him or her into your chest and then standing upright.

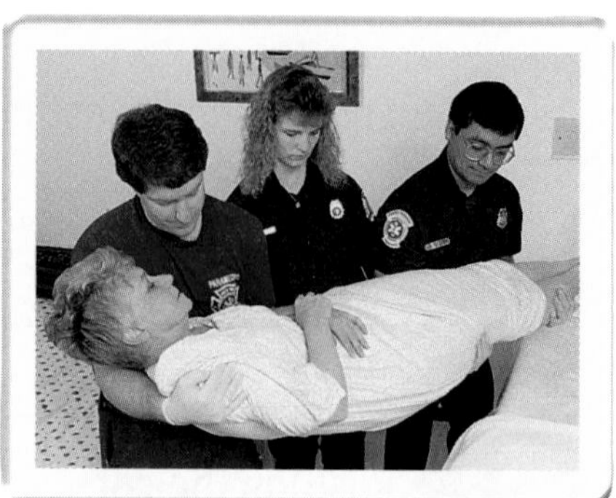

Step 5 Move and position patient.

Move the patient to the cot and lower patient carefully. Position as appropriate and secure.

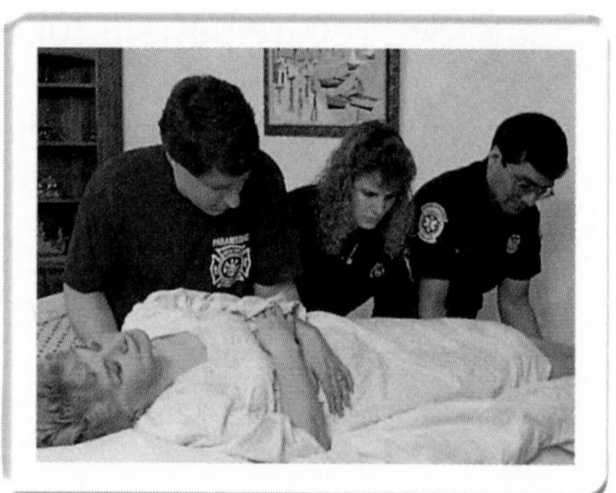

Infection Control Techniques

Performance Objective

Given standard patient contact situations, the candidate will demonstrate appropriate body substance isolation, disposal of contaminated sharps, and disposal of contaminated equipment and materials, in 1 minute or less.

Equipment

The following equipment is required to perform this skill:

- Appropriate body substance isolation/personal protective equipment
- Wraparound-style eye protection, designed to prevent splash behind the lenses
- Surgical mask
- HEPA filter mask
- Disposable gown
- Puncture-resistant sharps container
- Biohazard disposal box, with red bags

Indications

- Known cases of infectious patients
- Preparation for care of patients where infectious risk is unknown
- Clean-up following patient care

Contraindications

- None

Complications

- None

Procedures

Disposing of Contaminated Sharps

  **Ensure body substance isolation before beginning procedures.**

Prior to beginning patient care, appropriate body substance isolation procedures should be employed.

  **Place sharps in puncture-resistant container.**

As soon as possible following use, contaminated sharps should be put into a specifically designed, puncture-resistant container. Never shove a needle into the container. Do not recap the needle before placing it into the container.

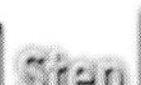  **Seal and dispose of puncture-resistant container.**

When the sharps container is two thirds to three fourths full, secure the cover. Place in a biohazard disposal box.

Disposing of Contaminated Equipment and Materials

Ensure body substance isolation before beginning procedures.

Prior to beginning patient care, appropriate body substance isolation procedures should be employed.

Place contaminated disposable equipment and materials in biohazard bag.

Place disposable equipment and materials in a specifically designed biohazard disposal bag.

Dispose of biohazard bag.

Place the full biohazard disposal bag into a biohazard disposal box.

Seal the biohazard box.

Seal the biohazard box *before* it becomes completely full.

Crime Scene Operations

Performance Objective

Given a patient found in an active crime scene, the candidate will demonstrate proper patient and evidence management in cooperation with law enforcement, in 2 minutes or less.

Equipment

The following equipment is required to perform this skill:

- Appropriate body substance isolation/personal protective equipment
- Paper bags

Indications

- Operations in an active crime scene

Contraindications

- None

Complications

- None when properly applied

Procedures

Working in the Crime Scene

 **Ensure body substance isolation before beginning procedures.**

Prior to beginning patient care, appropriate body substance isolation procedures should be employed.

 **Minimize contact with the scene.**

Everything surrounding the patient contains potential evidence. Touch or move only those items that are absolutely necessary. It is best to eliminate unnecessary equipment and personnel by placing them outside the actual crime scene. Even the cot can be placed outside the immediate area in most cases.

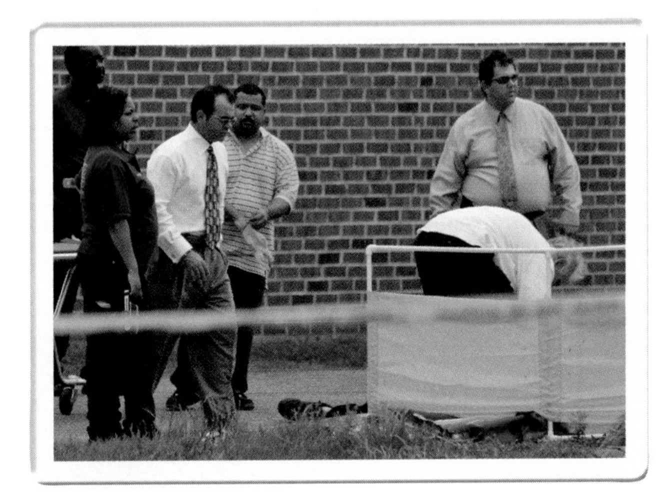

 **Minimize contamination.**

Be careful not to contaminate the scene with "ambulance droppings." As the time and the situation permits, place wrappings and packaging in appropriate containers. This should be performed as equipment and supplies are opened and before the packaging materials reach the ground. Any wrappings that reach the ground should be left to avoid gathering or destroying evidence that may exist underneath. Be careful about where you position equipment bags, monitors, and cots.

 **Document.**

After finishing patient care, document for law enforcement personnel everything in the crime scene that was moved or touched in the care of the patient. A sketch of the scene may be helpful. This should be performed on a report separate from the patient care record. It is helpful for the police investigation for you to give your name and contact information. This will allow them to contact you if any questions arise concerning the operation in the crime scene.

Preserving Evidence

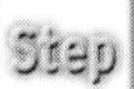 **Ensure body substance isolation before beginning procedures.**

Prior to beginning patient care, appropriate body substance isolation procedures should be employed.

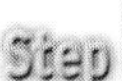 **Cut necessary clothing from the patient.**

Remove clothing by cutting carefully. Do not use stab holes, gunshot holes, or rips caused by assault as starting points for cutting. Do not shake the clothing in order to preserve trace evidence.

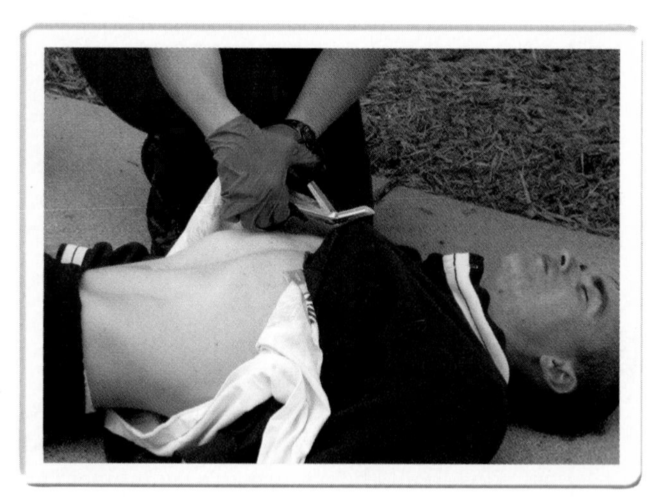

Step 3 Collect patient's clothing.

Place any clothing removed from the patient into individual paper bags. Placing pieces of clothing in the same bag or clothing in plastic bags will cause evidence to blend or sweat.

Protect the patient's hands.

Place paper bags over the patient's hands to protect evidence.

Use caution bandaging wounds.

Use caution cleaning up any wounds present on the patient. Bandaging and dressing should be limited to wounds that absolutely require bleeding control.

Managing a Violent Patient

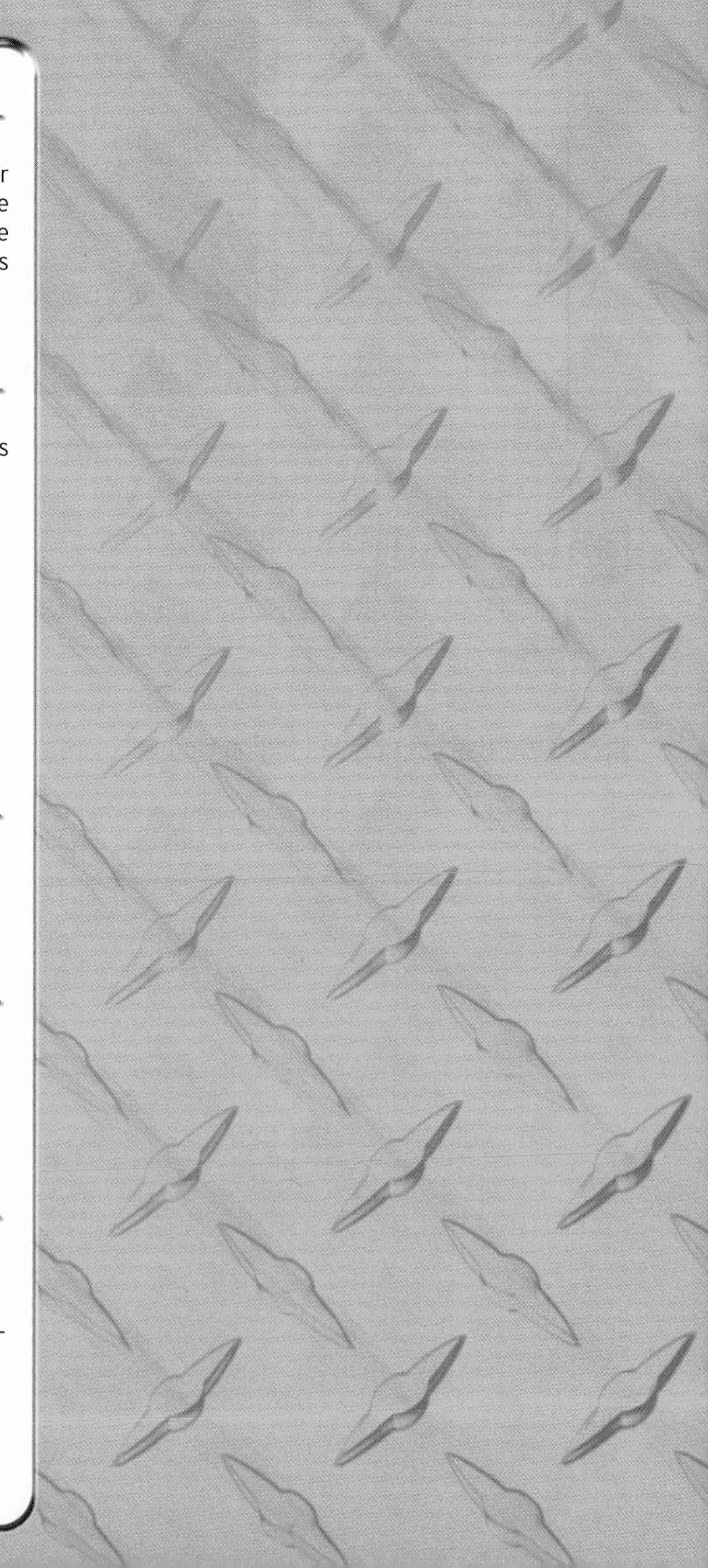

Performance Objective

Given a violent patient in danger of harming himself or herself or others, and working with other rescuers, the candidate(s) will demonstrate the proper procedure for subduing and restraining the patient, in 2 minutes or less.

Equipment

The following equipment is required to perform this skill:

- Appropriate body substance isolation/personal protective equipment
- Multi-level ambulance cot
- Commercial restraints for wrists and ankles
- Backboard straps

Equipment that may be helpful:

- Fluffy roller gauze
- Triangular bandages

Indications

- Violent patients who must be restrained for their own safety or the safety of others

Contraindications

- Insufficient personnel trained and prepared to physically restrain the patient
- Patients with weapons

Complications

- Injury to rescuers
- Injury to the patient
- Cardiac arrest from excited delirium or other metabolic states associated with stimulant intoxication (in-custody death syndrome)

Procedures

Step 1 Ensure body substance isolation before beginning procedures.

Prior to beginning patient care, appropriate body substance isolation procedures should be employed.

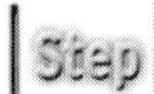 

Step 2 Request police and remove others from danger.

As soon as the danger has been identified, call for police assistance. Remove unnecessary personnel from the scene and maintain a safe distance.

Step 3 Attempt to calm the patient verbally.

Using a calm voice, talk to the patient and attempt to calm him or her. To avoid confusion, only one responder should communicate with the patient. Avoid threatening, derogatory, or demanding tones as you attempt to gain the patient's cooperation. You should start by asking questions to determine the cause of the bizarre and dangerous behavior. As you ask questions, avoid questions starting with "why." "Why" questions force patients to justify their actions. Our intent is not for patients to *justify* what they are doing, but instead to *identify* what they are doing. Be prepared to spend time talking to your patient. As long as the patient keeps talking and is not posing an immediate threat to himself, herself, or others, time is on your side.

Advise patient of the need for restraint and continue to give opportunity to comply.

If negotiation and conversation fail, it is necessary to give the patient specific instructions as to how to behave. These instructions should be specific, with a reasonable time frame in which to comply. Included in these instructions should be a warning that further actions will result in the need to be physically restrained.

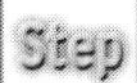 

Formulate plan and make assignments.

While the person performing the negotiations talks to the patient, other members of the patient care team should formulate plans for restraining the patient. These plans should include who will be responsible for controlling which body part. It is best for three to four people to attempt the restraint. Fewer people are often insufficient, and more people often get in the way. As you plan your actions, assign people to the following sites:

- One person should approach from behind to control the patient's legs.
- One person should be assigned to each arm and be prepared to approach from the sides.
- One person should approach from behind to grab the patient's head.
- Any additional personnel should be held in reserve to assist as needed.

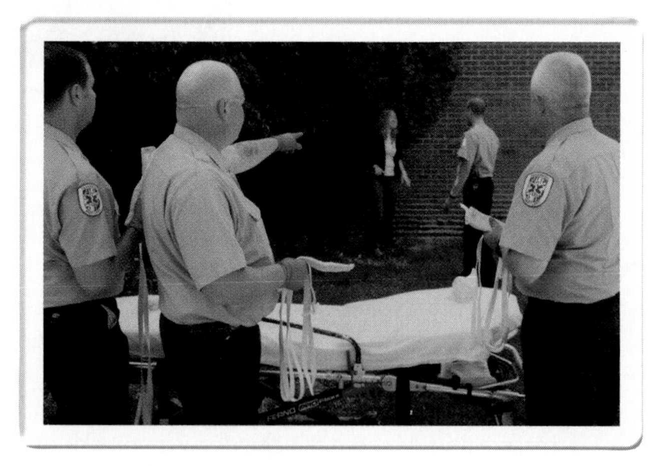

Step 6 Surround patient and give last opportunity to comply.

Get into the proper position to subdue the patient. The person doing the negotiation should continue to talk to the patient and attempt to get the patient to comply.

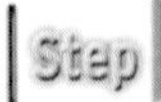

Step 7 Grab and control the patient.

Immediately after the patient has failed to comply when given his or her final opportunity, move forward and gain control. Remember the following rules as you approach and grab the patient:

- Act fast and act together to control the patient. Do not give the patient an opportunity to develop his or her own counterplan.
- Restrain each arm by grabbing the patient by the upper arm while controlling the lower arms. Grabbing the wrists give the patient room to wiggle free.
- Restrain the legs by grabbing the lower thighs together and pulling backward. Approaching from behind reduces the chances of being kicked. If the patient attempts to kick backward, the rescuers controlling the arms can use his or her altered balance to put the patient on the ground.
- Restrain the head to control the general movement of the body. Be careful not to get any part of your body in a position to be bitten. Also use caution not to strangle the patient by grabbing around the neck.

Step 8 Place the patient in a seated position on the cot.

There may be a brief period during which the patient is lying on the ground either prone or supine. This position should be extremely brief to prevent issues with positional asphyxia.

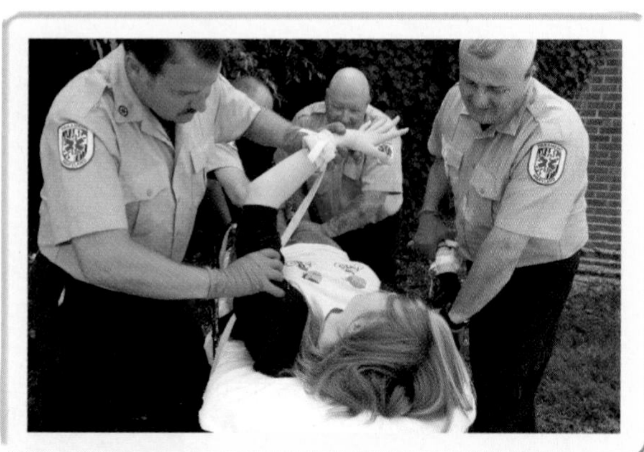

Once the patient is controlled, move the patient onto the cot.

Working in unison, put the patient in a seated position on the cot as soon as possible. Avoid the use of handcuffs before the patient is placed on the cot. Because the patient will need to have his or her hands secured with a wider and padded restraint, handcuffs will create more problems than they will solve.

Step 9 Secure extremities and the patient to the cot.

Apply straps to the patient's chest, hips, and thighs. Wrist restraints should be applied and tied to the sides of the cot. Ankle restraints may be applied if needed. Commercial restraints are best because they are wide and padded. If unavailable, triangular bandages or fluffy roller gauze can be made into a restraint as well. Be sure the restraints do not cut off the patient's distal circulation. If the patient begins spitting, place a barrier over his or her mouth.

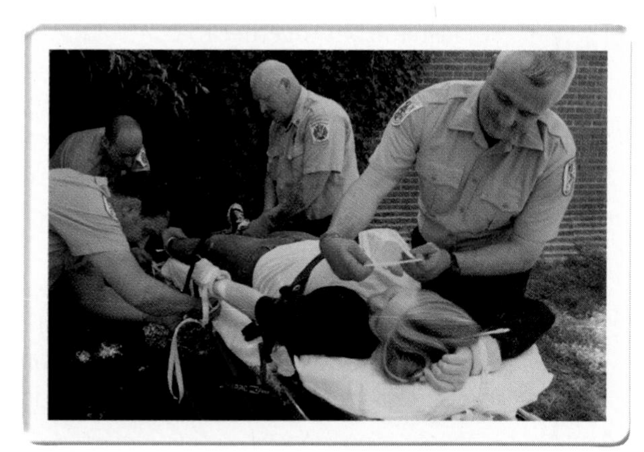

In the Field

Tasers

Police are now utilizing Tasers and other electrical stun devices as a means of controlling uncooperative and noncompliant suspects. These devices are in and of themselves nonlethal. However, there have been numerous cases in which patients have died shortly after the use of Tasers. In nearly all cases these deaths have resulted not from the Taser but from the stimulants that the patient had used before being tased. To ensure proper assessment and care of those who have been tased, it is important for police agencies and EMS to establish strict policies concerning the assessment of every patient.

Step 10 Assess and manage patient.

Once the patient is secured to the cot, perform the necessary physical examination. This may be as simple as identifying that the patient continues to breathe and is conscious. Any further assessment may simply raise the patient's aggressiveness and force him or her to fight the restraint.

Patients who lose consciousness during restraint need immediate assistance. Assess the patient's temperature and manage hyperthermia accordingly.

Check Your Knowledge

You are dispatched to a local nursing home to transport a patient to an acute care facility. You arrive on scene to find a frail looking 86-year-old woman lying supine in bed. The nursing staff tells you that the patient's doctor has ordered that she be admitted to the hospital for a general decline in condition. They tell you that the patient is non-ambulatory and is not able to communicate. The patient is on continuous oxygen via nasal cannula. Your initial assessment reveals that the patient does not appear to be in any distress. She is breathing adequately and her vital signs are stable.

1. **Which of the following would be the best method for transferring this patient from her bed to the cot?**
   A. Two-person extremity lift
   B. Sheet technique
   C. Fireman's carry
   D. Secured on a long backboard
2. **When transporting the patient on the cot, why is it recommended that you position the cot no higher than position three, four, or five?**
   A. Higher positions may be unstable.
   B. It is easier to move the cot.
   C. It is easier to place the cot into the ambulance.
   D. The patient may become afraid at higher positions.
3. **What is the best position for transport of this patient?**
   A. Trendelenberg position
   B. Fowler's position
   C. Recovery position
   D. Position of comfort
4. **If it is necessary to descend stairs with the patient on the cot, how should you move the patient?**
   A. Feet first
   B. Head first
   C. Sitting up
   D. None of the above
5. **Proper body mechanics for lifting a patient include which of the following?**
   A. Spread feet about shoulder-width apart.
   B. Squat into position rather than bending over.
   C. Keep your back straight.
   D. All of the above.

Additional Questions

6. **Proper disposal of contaminated sharps includes which of the following?**
   A. Recap needles before disposal.
   B. Dispose of sharps in a puncture-resistant sharps container.
   C. Leave all sharps at the hospital.
   D. Clean sharps before disposal.
7. **Which of the following statements is true with regard to working a potential crime scene?**
   A. Preservation of evidence is not the responsibility of EMS personnel.
   B. You may not share information with law enforcement officials.
   C. Touch or move items only when absolutely necessary.
   D. You should be sure to pick up and dispose of any discarded wrappings you may drop during patient care.
8. **When working on a patient at a potential crime scene, it is important that you:**
   A. Use existing holes in the patient's clothing as a starting point if you need to cut clothing.
   B. Clean all wounds thoroughly.
   C. Place any clothing removed from the patient in individual paper bags.
   D. All of the above.
9. **When managing a violent patient, what should be your primary concern?**
   A. Your safety
   B. Restraining the patient
   C. The patient's safety
   D. Protecting bystanders
10. **The steps for restraining a violent patient include which of the following?**
   A. Advise and instruct the patient.
   B. Have a plan and make assignments.
   C. Place a barrier device over the patient's mouth if he or she begins spitting.
   D. All of the above.

Advanced Airway Management

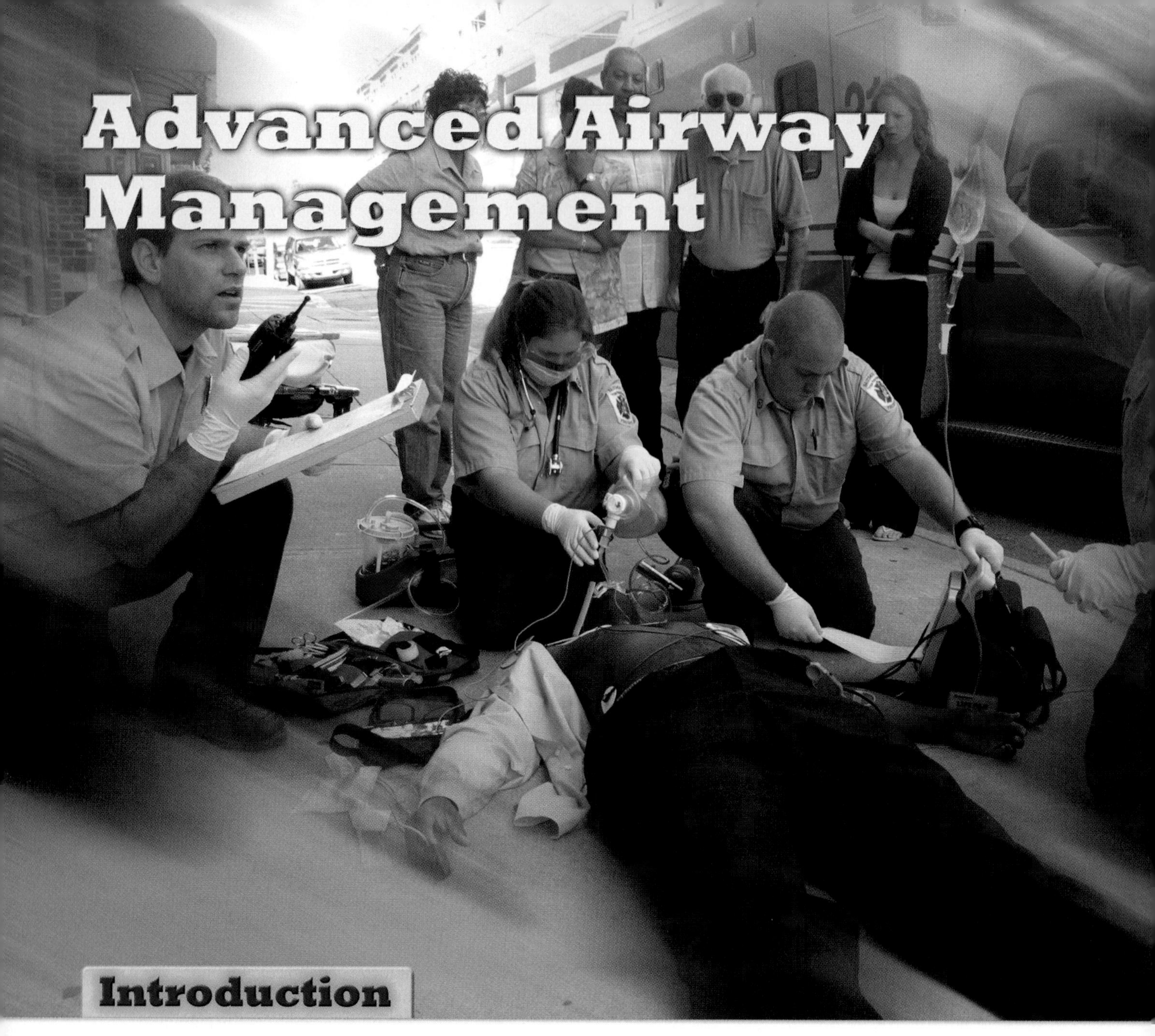

Introduction

Airway management is essential to patient care. Advanced airway management takes the process to a higher level, with higher responsibilities. Properly placed, the advanced airway helps to ensure the ventilatory process of the patient. Improperly placed and uncorrected, the advanced airway can lead to fatal consequences for the patient and for the responder's career.

Many EMS systems and many states are moving to the use of simpler advanced airways as a basic life support skill. It is important to remember that advanced airways are simply adjuncts to secure and maintain the patient's airway. They assist in the delivery of oxygenation and ventilation by ensuring a direct route through the larynx and into the lungs. Mastering advanced airway skills takes constant, daily practice to ensure proper placement under emergency situations. The skills in this section will guide the responder through the process of securing and maintaining an airway. Be sure to check local protocol before performing an advanced procedure.

SECTION 8

SKILL 46

Dual-Lumen Airway Devices

Performance Objective

Given an adult patient and appropriate equipment, the candidate shall insert a dual-lumen airway and verify proper placement using criteria herein prescribed, in 6 minutes or less.

Equipment

The following equipment is required to perform this skill:

- Appropriate body substance isolation/personal protective equipment
- Oxygen cylinder, regulator, and key
- Oxygen delivery device (appropriate to patient)
  - Bag-mask device
- Oropharyngeal airway (adult sizes)
- Water-soluble lubricant
- Dual-lumen airway
  - Combitube
  - KING LT-D*
  - Pharyngeotracheal lumen (PtL) airway
- Syringes (sizes 20 mL to 100 mL; usually included in kit with airway)
- Tube holder or tape
- Stethoscope
- Suction device
- Suction tubing
- Suction catheter
  - Rigid wand (Yankauer type)
  - Flexible catheter

Equipment that may be helpful:

- Pulse oximeter
- End-tidal carbon dioxide meter

*The KING LT-D airway is not dual lumen. However, it is included here because it is an alternative airway device with similar placement and procedures to true dual-lumen airways.

Indications

- Failed attempt at endotracheal intubation or difficult anatomy
- Confined-space airway control
- Advanced airway care not immediately available

Contraindications

- Intact gag reflex
- Esophageal disease
- Ingestion of caustic (acid/alkali) substance
- Patients with laryngeal edema from anaphylaxis, respiratory burns, or other causes

Complications

- Sore throat
- Dysphagia
- Esophageal hemorrhage/rupture
- Upper airway bleeding
- Laryngeal occlusion (if airway is placed too deep)

Procedures

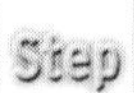

Step 1 Ensure body substance isolation before beginning procedures.

Prior to beginning patient care, appropriate body substance isolation procedures should be employed.

Step 2 Open the airway manually.

Before opening the airway, consider the possibility of cervical spine injury. If spinal injury is suspected, use a jaw-thrust maneuver to open the airway.

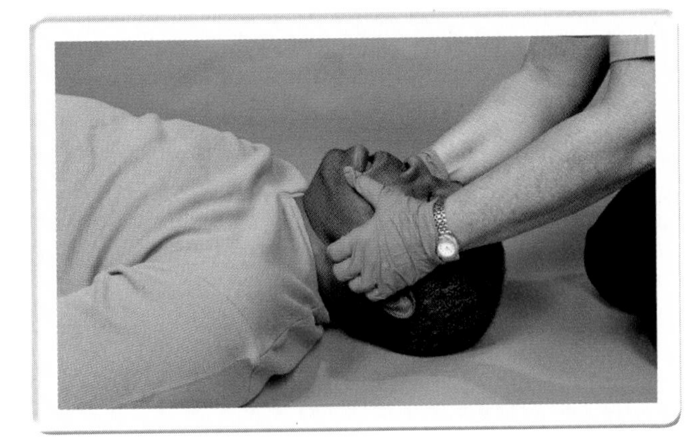

If no spinal cord injury is suspected, open the patient's airway by performing the head tilt–chin lift procedure.

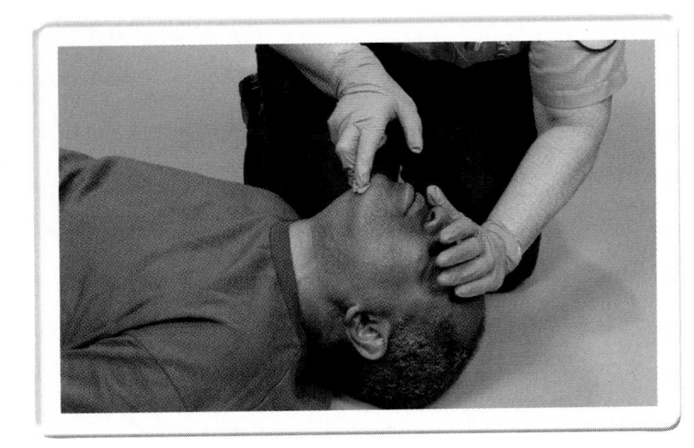

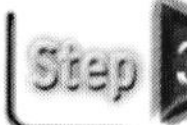

Step 3 Elevate tongue and appropriately insert simple adjunct (see Skill 7).

Measuring from the corner of the mouth to the base of the earlobe, choose the correct-sized airway for your patient. Open the mouth using the jaw-thrust or cross-finger technique. Insert the airway in the front of the mouth with the tip pointed toward the roof, or insert from the side of the mouth with the tip toward the inside of the cheek. A tongue depressor can be used to hold the tongue inferior and the airway inserted tip down. If the patient starts to gag, remove the airway by pulling the flange anterior and inferior. If gag reflex remains, dual-lumen placement is contraindicated.

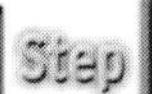

Step 4 Ventilate the patient immediately with bag-mask device.

Select the appropriate-sized bag for the patient. Bag and mask sizes are normally listed as adult, child, and infant. These sizes are adequate for the average-sized person meeting the age definition criteria. However, large children will require a bag that will deliver the appropriate volume. Likewise, pediatric bags may be appropriate for small adults or adults in whom overpressurization could cause pulmonary damage. Masks should be chosen to fit the patient without air leakage.

Position the mask with the apex over the bridge of the nose and the base between the lower lip and the prominence of the chin. Begin ventilations as soon as the mask is sealed, and assess for air leakage.

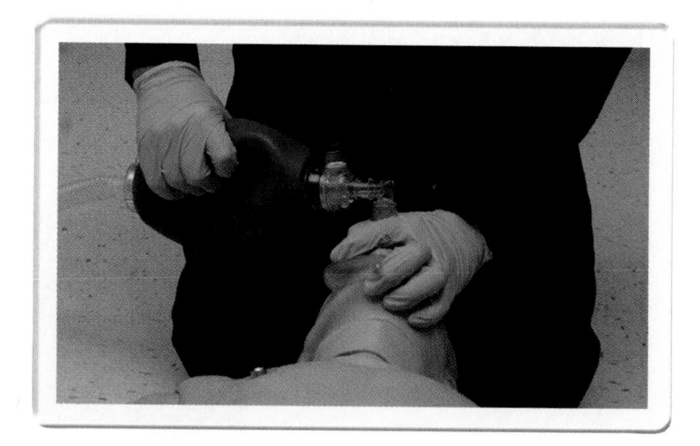

The initial ventilations of a nonbreathing patient should be performed in a controlled and accelerated manner. Deliver 24 breaths/min for 2 to 3 minutes. Use extreme caution not to overventilate. Each breath should be given slowly, with adequate time for exhalation in between.

After 2 to 3 minutes of preoxygenation, deliver ventilations at a rate of 10 to 12 breaths/min, with a tidal volume of at least 800 mL. Pay close attention to ensure that each ventilation is of the appropriate volume and of a consistent rate. Allow adequate exhalation between each breath.

continued

Effective ventilations are those breaths that cause the chest to rise adequately for the size of the patient. In addition to assessing chest rise, consider checking breath sounds. Be cautious not to overventilate. Each ventilation should be delivered slowly and easily, lasting 1 second. Fast and high-pressure ventilations can cause air to enter the stomach, increasing the risk of vomiting. If a second rescuer is available, the Sellick maneuver (cricoid pressure) may be applied. Although it will be necessary to stop ventilations for assessment and other procedures, you should never allow a nonbreathing patient to go without ventilations for longer than 30 seconds.

Provide supplemental oxygen.

If not previously performed, attach oxygen tubing to the inlet on the bag-mask device and to the oxygen source. Open the cylinder valve and adjust liter flow to between 12 and 15 L/min (see Skill 4).

This skill describes steps for placement and removal of three different types of airways: Combitube, PtL, and KING LT-D. The steps for the placement and removal of the Combitube begin below. The steps for placement and removal of the PtL begin on page 275 and the steps for the KING LT-D begin on page 279. Be sure to complete Steps 1 through 5 first.

Placement of Combitube

Direct assistant to take over ventilation.

Direct a qualified assistant to take over ventilations.

Step 7 Select, check, and prepare the Combitube.

While the patient is being ventilated, the person responsible for the intubation must ensure that all the necessary equipment is available and in working order. Two sizes of Combitube are available. Check to ensure that the correct tube is utilized from the following chart. Each tube will require a different amount of air in the esophageal cuffs. Choose the correct-sized Combitube for the patient.

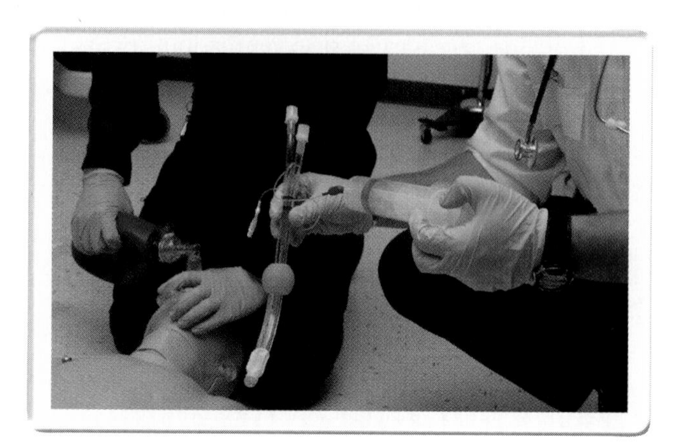

	Combitube	Combitube SA (small adult)
Intended patient heights	5′ and taller (caution over 6′ 8″)	4′ to 5′ 6″
Pharyngeal cuff	100 mL	85 mL
Esophageal cuff	15 mL	12 mL

Each kit comes complete with appropriately sized syringes, preloaded with the correct amount of air. Check these to ensure that the proper level has been set, and test the inflation of each cuff by inflating to the level indicated on the color-coded inflation ports. Refill each syringe to the proper volume of air. The Combitube kit also includes a nonsterile 10F suction catheter and a fluid deflector elbow. It is advisable to place the deflector on the end of the white tube before inserting the airway, aimed to deflect fluids toward the patient's side and away from active rescuers.

Additionally, always ensure that suction is available and working before beginning intubation. Both a rigid suction tip and a flexible catheter, appropriate for the size of the endotracheal tube, should be available.

Step 8 Lubricate distal tip of the Combitube.

The Combitube will pass in most patients with the natural lubrication of oral fluids. However, in patients who are dry, adding a water-soluble lubricant may be helpful.

Position patient's head properly.

Have your assistant stop ventilations, remove the oral airway, and remove the mask from the bag-mask device. You will now have only 30 seconds to position the patient, place the Combitube airway, and reestablish ventilations.

Should you fail to intubate the patient, your assistant should quickly place the mask back on the bag-mask device and begin hyperventilating the patient.

Gently return the patient's neck to a neutral or slightly flexed position.

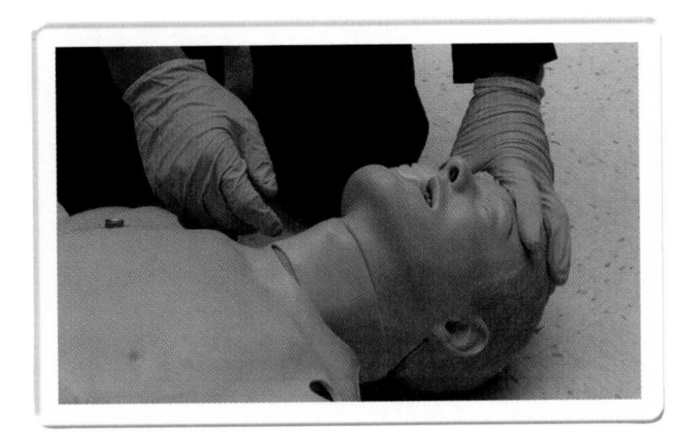

Insert the Combitube midline and to a depth such that the printed ring is at the level of the teeth.

Place your thumb into the patient's mouth and lift the tongue and lower jaw anterior and inferior (tongue–jaw lift maneuver). Insert the Combitube and gingerly guide it straight down the center of the oral cavity into the hypopharynx. Stop advancing the tube when the patient's teeth are between the two black reference lines.

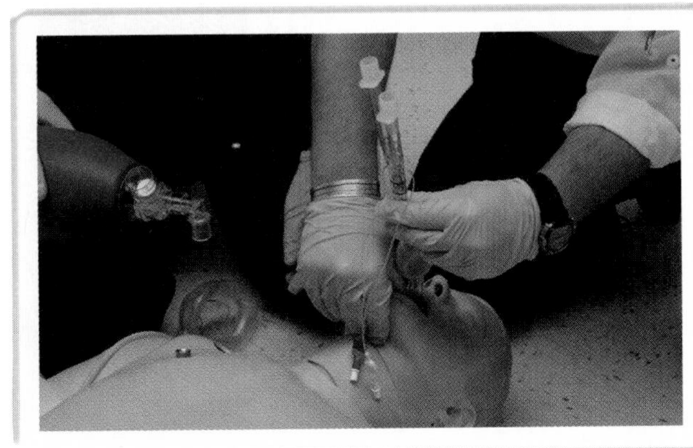

Step 11 Inflate the pharyngeal cuff with the proper volume and remove syringe.

Inflate the pharyngeal cuff first through the *blue* pilot balloon port, using the prescribed amount of air (100 mL for Combitube, 85 mL for Combitube SA). Remove syringe and check pressure in the pilot balloon.

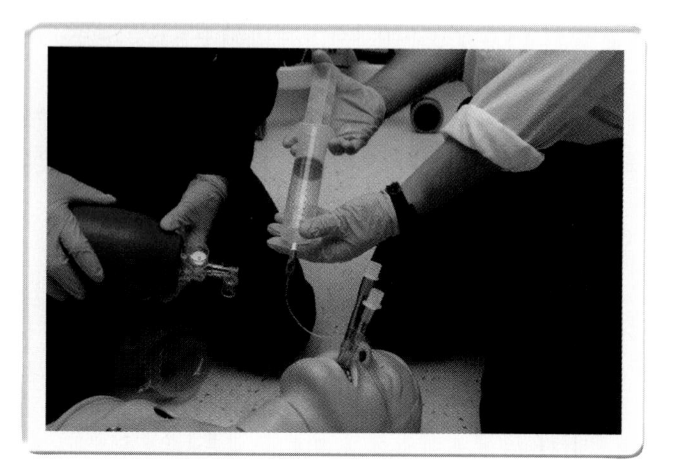

Step 12 Inflate the distal cuff with the proper volume and remove the syringe.

Inflate the esophageal cuff next through the *white* pilot balloon port, using the prescribed amount of air (15 mL for Combitube, 12 mL for Combitube SA). Remove syringe and check pressure in the pilot balloon. Should either balloon show signs of insufficient volume, add air as needed.

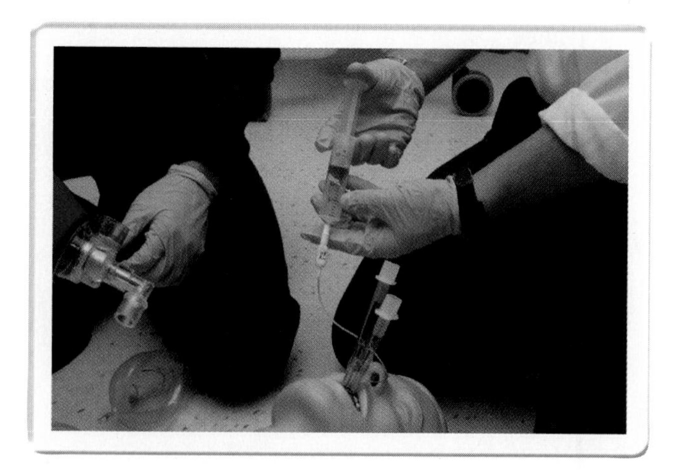

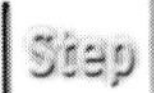

Step 13 Attach or direct attachment of bag-mask to the first (pharyngeal placement) lumen and ventilate.

Attach bag-mask device to the *blue* tube and assess position by listening to abdominal and breath sounds.

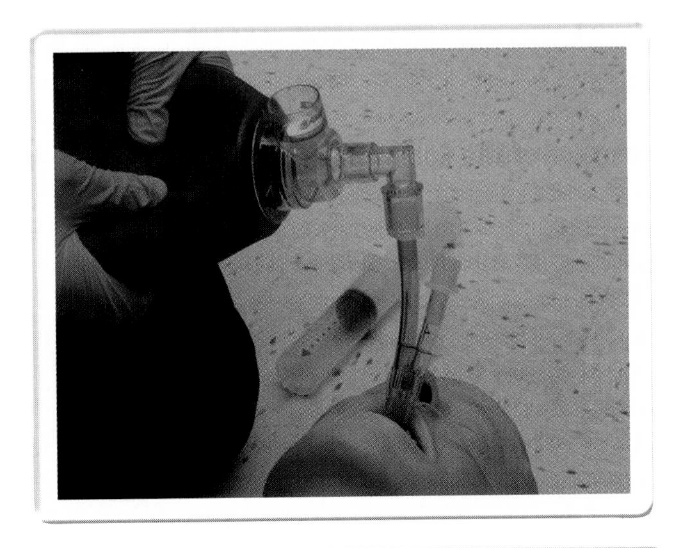

Step 14 Confirm placement and ventilation through correct lumen by observing chest rise, auscultation over the epigastrium, and auscultation bilaterally over each lung.

Listen first over the patient's gastric region. With absent abdominal sounds, listen over the lung fields to assess the quality of ventilation.

The presence of abdominal sounds is an indication of tracheal placement of the Combitube. Remove the fluid diverter elbow from the *white* tube, attach the bag-mask device, and reassess abdominal and breath sounds.

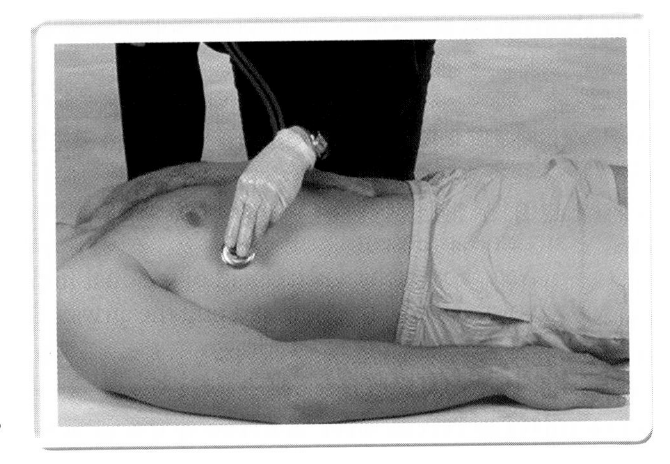

If sounds are not heard in either location, reposition the Combitube slightly before complete removal. To reposition the tube, deflate both cuffs and pull the Combitube back 2 to 3 cm, reinflate the cuffs, and reassess. If sounds are still not heard, remove the Combitube and consider alternative airway procedures.

Step 9 Position patient's head properly.

Have your assistant stop ventilations, remove the oral airway, and remove the mask from the bag-mask device. You will now have only 30 seconds to position the patient, place the esophageal airway, and reestablish ventilations. Should you fail to intubate the patient, your assistant should quickly place the mask back on the bag-mask device and begin hyperventilating the patient.

Gently return the patient's neck to a neutral or slightly flexed position.

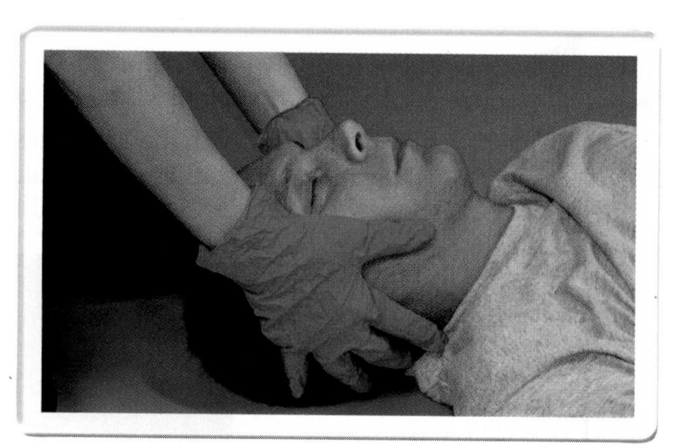

Step 10 Insert PtL midline and to a depth such that the strap is at the level of the teeth.

Place your thumb into the patient's mouth and lift the tongue and lower jaw anterior and inferior (tongue–jaw lift maneuver). Insert the PtL so the curvature of the tube is aligned with the natural curvature of the airway.

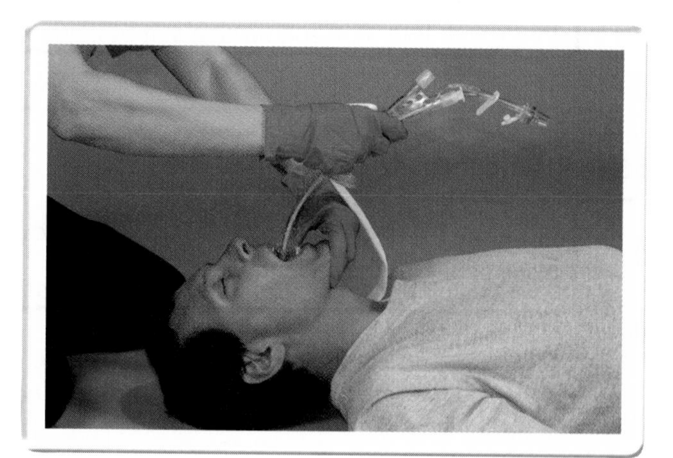

Apply gentle downward pressure to advance the tube until the teeth strap touches the patient's teeth or gums. Slight resistance is expected as the tube passes the back of the oropharynx. *Do not* force the airway. If the PtL does not advance easily, either redirect the tube by slightly adjusting the airway or withdraw the tube and start over.

Step 11 Secure PtL.

Position the neck strap around the patient's neck and attach it to the tube using the fastening system.

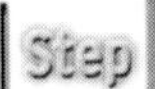

Step 12 Inflate cuffs with proper volume.

Inflate the cuffs of the PtL by placing the ventilation port of a bag-mask device against the inflation valve (labeled tube 1). Ensure that the white cap is closed to prevent leakage.

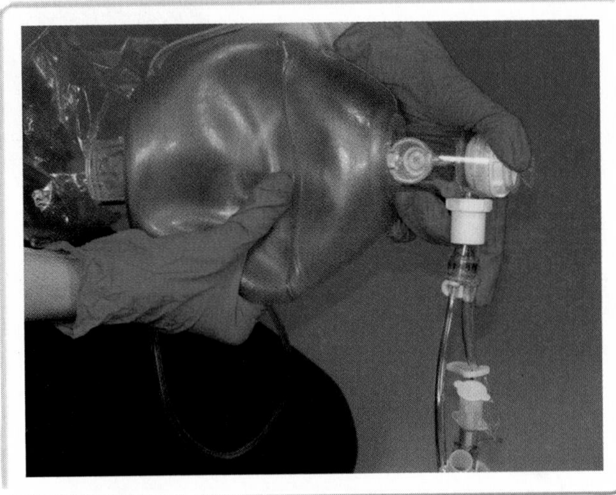

Step 13 Attach or direct attachment of the bag-mask device to tube 2 (the short, green tube) and assess placement.

Attach the bag-mask device to the short, green tube (labeled tube 2) and assess position by observing chest rise and listening to abdominal and breath sounds. If the chest rises, tube 3 is in the esophagus. Continue ventilating through tube 2.

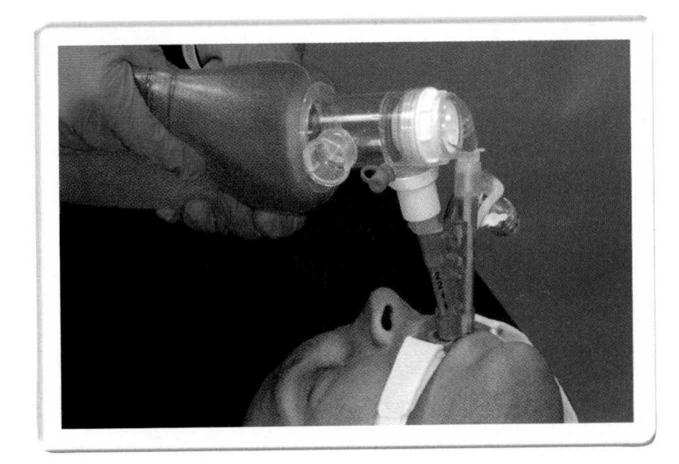

 continued

If the chest does not rise, tube 3 may be in the trachea. Remove the stylet from the long, clear tube (labeled tube 3). Attach the bag-mask device and ventilate through the clear tube. Assess position by observing chest rise and listening to abdominal and breath sounds.

Continue ventilation through the appropriate tube.

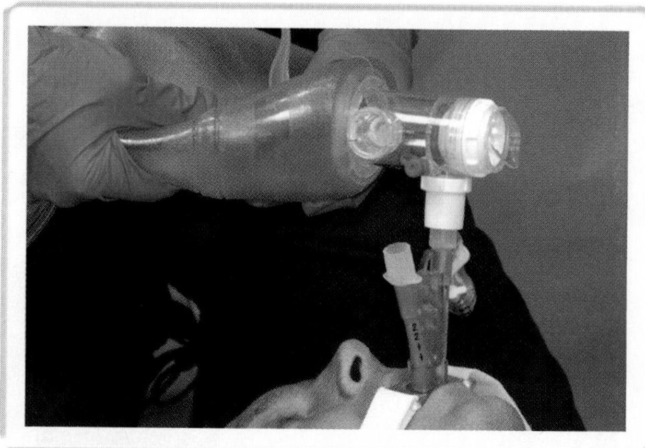

Removal of the PtL (From Esophageal Placement)

 **Deflate both cuffs.**

Open the white cap on tube number 1 and allow air to escape from both tubes.

 **Pull PtL out.**

With suction ready and the patient's head in a neutral position, pull the PtL straight out.

 **Turn patient's head to side and suction.**

Turn the patient's head to the side (unless contraindicated by trauma) and suction any emesis present.

Placement of KING LT-D

The KING LT-D is a single-lumen airway. Unlike the Combitube, PtL, or other dual-lumen airways, the KING LT-D is only effective when placed in the esophagus. Because the KING LT-D is not a dual-lumen airway, inadvertent intubation of the trachea must be recognized and corrected as soon as possible. Failure to recognize tracheal intubation can have fatal consequences.

Step 6 Direct assistant to take over ventilation.

Direct a qualified assistant to take over ventilations.

Step 7 Check and prepare KING LT-D.

While the patient is being ventilated, the person responsible for the intubation must ensure that all the necessary equipment is available and in working order. Three sizes of KING airways are available. Check to ensure that the correct tube is utilized from the following chart. Each tube will require a different amount of air in the esophageal cuffs.

Choose the correct-sized KING airway for the patient.

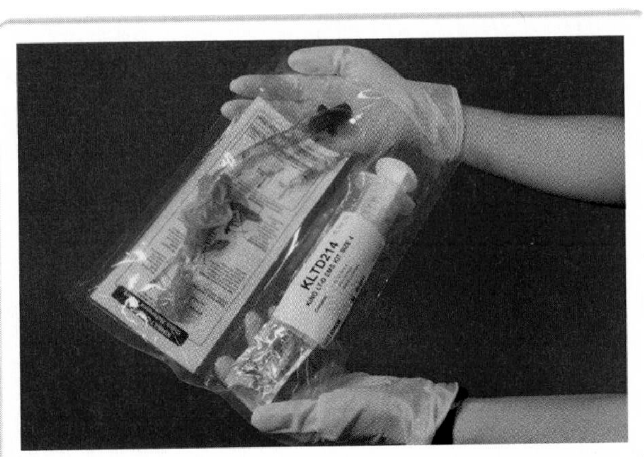

	KING LT-D Size 3	KING LT-D Size 4	KING LT-D Size 5
Intended patient heights	4′ to 5′	5′ to 6′	> 6′
Connector color	Yellow	Red	Purple
Inflation volume	45-60 mL	60-80 mL	70-90 mL

Test the cuff and the inflation system for leaks by placing the maximum recommended volume into the cuff. Before inserting, remove all air from the cuffs.

The manufacturer advises that you have a spare KING LT-D ready and prepared for immediate use.

Additionally, always ensure that suction is available and working before beginning intubation. Both a rigid suction tip and a flexible catheter, appropriate for the size of the tube, should be available.

Step 8 Lubricate distal tip of the KING LT-D.

Using water-soluble jelly, lubricate the distal tip and posterior aspects of the tube. *Do not* apply lubricant in or near the ventilatory openings.

Step 9 Position patient's head properly.

Have your assistant stop ventilations, remove the oral airway, and remove the mask from the bag-mask device. You will now have only 30 seconds to position the patient, place the esophageal airway, and reestablish ventilations. Should you fail to intubate the patient, your assistant should quickly place the mask back on the bag-mask device and begin hyperventilating the patient.

Gently return the patient's neck to a neutral or slightly flexed position.

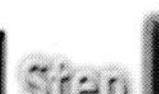

Step 10 Insert the KING LT-D midline and to a depth at which the base of the connector is properly aligned.

Place your thumb into the patient's mouth and lift the tongue and lower jaw anterior and inferior (tongue–jaw lift maneuver). With the KING LT-D rotated so the blue line is touching the corner of the mouth, insert the tip into the mouth and behind the tongue. Rotate the tube as the tip passes under the tongue so that the blue line faces the chin. Advance the tube until the base of the connector is aligned with the teeth or gums.

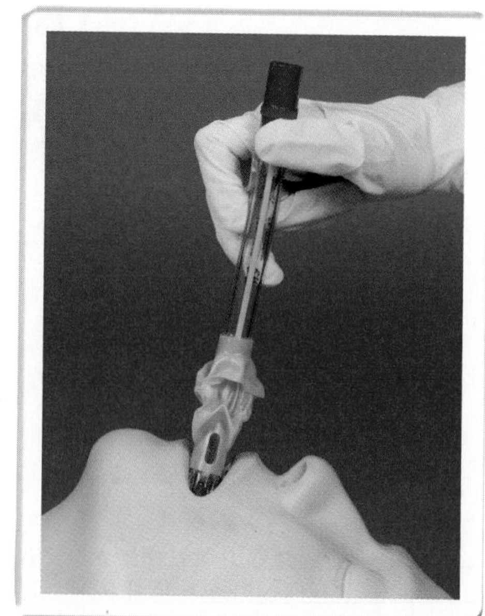

Step 11 Inflate cuffs with proper volume and remove syringe.

Using the syringe provided in the KING LT-D kit, inflate the cuffs with the appropriate volume: 50 mL for size 3, 70 mL for size 4, and 80 mL for size 5.

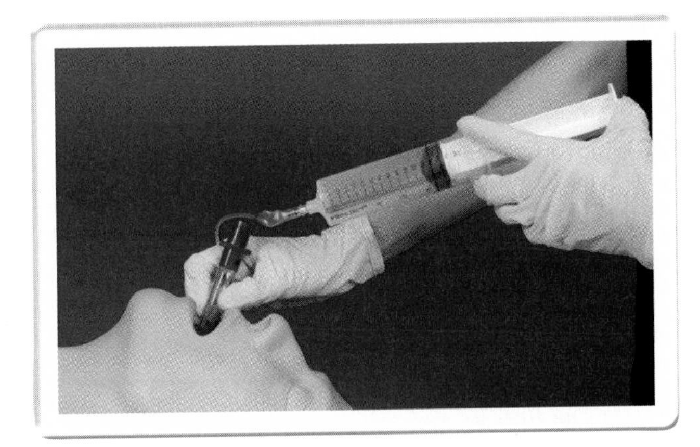

Remove the syringe from the inflation port.

Step 12 Attach or direct the attachment of the bag-mask device to the airway lumen and ventilate.

Attach the bag-mask device to the 15-mm connector and begin gentle ventilation.

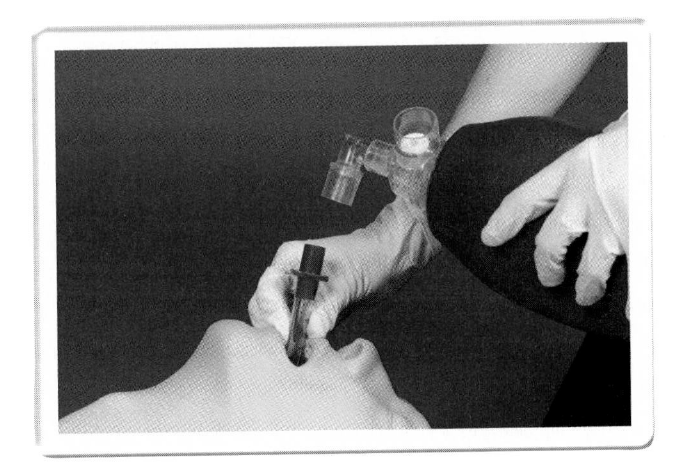

Assess breath and abdominal sounds while simultaneously pulling back on the KING LT-D until ventilation becomes easy. Withdrawing the tube with cuffs inflated retracts the surrounding tissues and aids in the securing of a patent airway. Ventilations should deliver large tidal volumes with minimal airway pressure when properly positioned.

Step 13 Secure the KING LT-D and confirm that the device remains properly secured.

Secure the KING LT-D using tape or a commercial tube holder. If tape is used, consider inserting an oropharyngeal airway as a bite block.

Removal of the KING LT-D (From Esophageal Placement)

Step 14 Prepare for removal.

Suction equipment should be present and in working order.

Step 15 Deflate cuffs.

Remove all air from both cuffs. It may take more than one filling of the syringe to completely remove the air.

Step 16 Pull KING LT-D out.

With suction ready and the patient's head in a neutral position, pull the KING LT-D straight out.

Step 17 Turn patient's head to side and suction.

Turn the patient's head to the side (unless contraindicated by trauma) and suction any emesis present.

Adult Endotracheal Intubation

SKILL 47

Performance Objective

Given an adult patient and appropriate equipment, the candidates shall insert an endotracheal (ET) tube and verify proper placement using the criteria herein prescribed, in 6 minutes or less.

Equipment

The following equipment is required to perform this skill:

- Appropriate body substance isolation/personal protective equipment
- Oxygen cylinder, regulator, and key
- Oxygen delivery device (appropriate to patient)
  - Bag-mask device
- Oropharyngeal airway (adult sizes)
- Laryngoscope handle and blades (straight and curved)
- Endotracheal tubes, cuffed (6.0 to 9.0)
- Malleable intubation stylet (appropriate size for ET tube)
- Water-soluble lubricant
- Towel or other padding
- End-tidal carbon dioxide meter
- Syringes (10 mL, 20 mL)
- ET tube holder or tape
- Stethoscope
- Suction device and tubing
- Suction catheter
  - Rigid wand (Yankauer type)
  - Flexible catheter

Equipment that may be helpful:

- Pulse oximeter
- Spare bulb, batteries
- Colorimetric end-tidal carbon dioxide detector

Indications

- Ensure definitive airway
- Assist with long-term ventilation

Contraindications

- None

Complications

- Unrecognized and uncorrected esophageal intubation
- Right mainstem bronchial intubation
- Laryngeal injury
- Dental injury

Procedures

Ensure body substance isolation before beginning procedures.

Prior to beginning patient care, appropriate body substance isolation procedures should be employed.

Open the airway manually.

Before opening the airway, consider the possibility of cervical spine injury. If spinal injury is suspected, use a jaw-thrust maneuver to open the airway.

If no spinal cord injury is suspected, open the patient's airway by performing the head tilt–chin lift maneuver.

Elevate the patient's tongue and appropriately insert simple adjunct (see Skill 7).

Measuring from the corner of the mouth to the base of the earlobe, choose the correct-sized airway for your patient.

Open the mouth using jaw-thrust or cross-finger technique. Insert the airway in the front of the mouth with its tip pointed toward the roof of the mouth, or insert it from the side of mouth with its tip toward the inside of the cheek.

A tongue depressor can be used to hold the tongue inferior and the airway inserted tip down. If the patient starts to gag, remove the airway by pulling the flange anterior and inferior.

Step 4 Ventilate the patient immediately with bag-mask device.

Select the appropriate-sized bag for the patient. Bag and mask sizes are normally listed as adult, child, and infant. These sizes are adequate for the average-sized person meeting the age definition criteria. Masks should be chosen to fit the patient without air leakage.

Position the mask with the apex over the bridge of the nose and the base between the lower lip and the prominence of the chin.

Begin ventilations as soon as the mask is sealed, and assess for air leakage.

The initial ventilations of a nonbreathing patient should be performed in a controlled and accelerated manner. Deliver 24 breaths/min for 2 to 3 minutes. Use extreme caution not to overventilate. Each breath should be given slowly, with adequate time for exhalation in between.

After 2 to 3 minutes of preoxygenation, deliver ventilations at a rate of 10 to 12 breaths/min, with a tidal volume of at least 800 mL. Pay close attention to ensure that each ventilation is of the appropriate volume and of a consistent rate. Allow adequate exhalation between each breath.

Effective ventilations are those breaths that cause the chest to rise adequately for the size of the patient. In addition to assessing chest rise, consider checking breath sounds. Be cautious not to overventilate. Each ventilation should be delivered slowly and easily, lasting 1 second. Fast and high-pressure ventilations can cause air to enter the stomach, increasing the risk of vomiting. If a second rescuer is available, the Sellick maneuver (cricoid pressure) may be applied.

Although it will be necessary to stop ventilations for assessment and other procedures, you should never allow a nonbreathing patient to go without ventilations for longer than 30 seconds.

Step 5 Provide supplemental oxygen.

If not previously performed, attach oxygen tubing to the inlet on the bag-mask device and to the oxygen source. Open the cylinder valve and adjust liter flow to between 12 and 15 L/min (see Skill 4).

Step 6 Direct assistant to take over ventilation.

Direct a qualified assistant to take over ventilations.

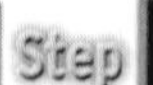 

Step 7 Select, check, and prepare the equipment.

While the patient is being hyperventilated, the person responsible for the intubation must ensure that all the necessary equipment is available and in working order. Ensure that the ET tube is the proper size for the patient.

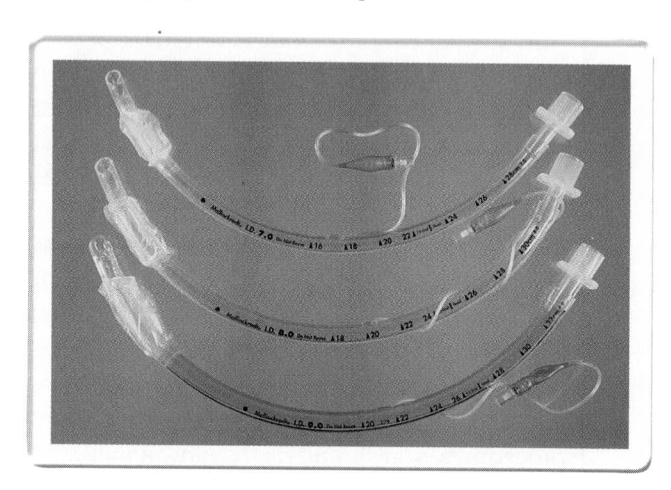

Using an aseptic technique, insert an intubation stylet into the ET tube and bend into a gentle curve. A properly curved stylet will allow the stylet to be removed without causing the tip of the tube to move. The stylet should be positioned so that the end of the stylet does not extend past the Murphy's eye.

Assemble the laryngoscope with the desired blade and check the brightness of the bulb. A bright white light is best for all intubations. A yellow or dull light will cause the tissues of the hypopharynx to take on similar color schemes and result in difficulties visualizing landmarks and the vocal cords. Make sure the light is turned off until the laryngoscope is placed in the patient's mouth.

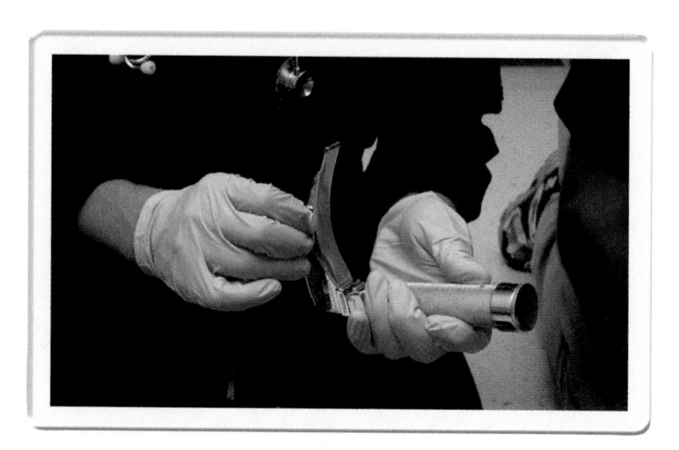

In the Field

Curved or Straight: Which Blade Is Best?

There seems to be a lot of discussion and conflict concerning which laryngoscope blade is the "best" blade to use. Both straight and curved blades have different advantages. It is best for each person to practice with both types of blades in order to develop skills and abilities rather than preferences. Resist the temptation to use only one type of blade for all intubations, and don't hesitate to switch blades should the need arise.

continued

Prepare for the intubation by placing a 1″ pad under the patient's head.

This will assist in lining up the three planes of the patient's airway. Although this step will make intubation easier, it is not required and is often difficult to accomplish in the pre-hospital setting. It is a good practice to learn to intubate both with and without the 1″ pad.

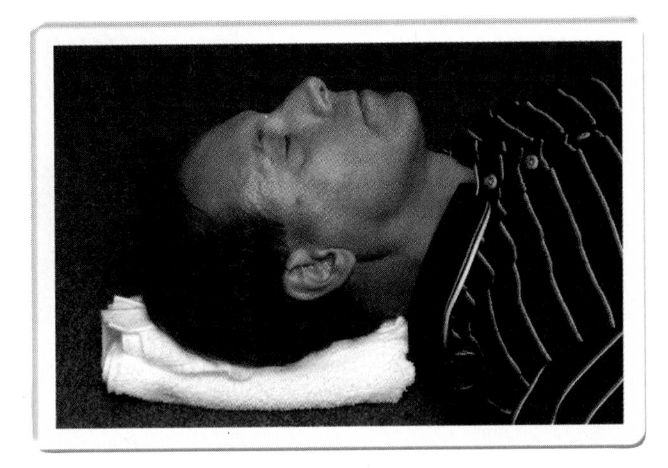

Finally, always ensure that suction is available and working before beginning intubation. Both a rigid suction tip and a flexible catheter, appropriate for the size of the endotracheal tube, should be available.

Position patient's head properly.

Have your assistant stop ventilations and remove the mask from the bag-mask device. You will now have only 20 seconds to position the patient, place the endotracheal tube, and reestablish ventilations. Should you fail to intubate the patient, your assistant should quickly place the mask back on the bag-mask device and begin hyperventilating the patient.

Remove the oropharyngeal airway and place it next to the patient's head or in another readily available location. It will be necessary to replace this airway upon successful intubation or in the event the patient is unable to be intubated.

Gently extend the patient's neck so that the head is positioned with the chin and forehead approximately 45° from the floor.

Step 9 Insert blade while displacing tongue, and visualize larynx.

Carefully insert the blade of the laryngoscope into the patient's mouth to the depth of the uvula. Using a sweeping motion, move the patient's tongue to the left as you lift at a 45° angle along the facial plane, keeping the laryngoscope off the teeth.

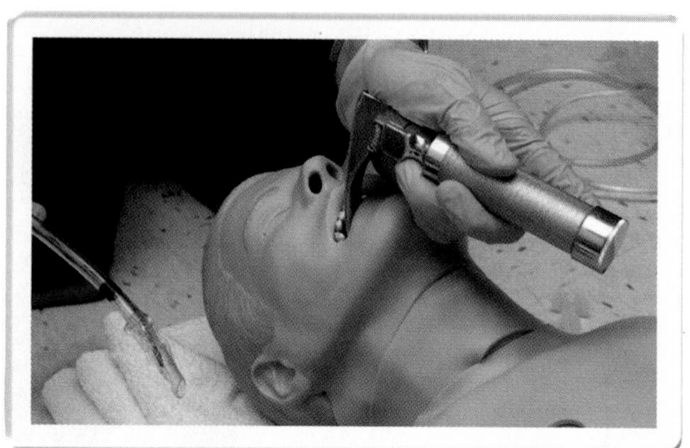

Lifting should be performed with a free hand. The habit of resting or placing the left forearm against the patient's forehead frequently causes prying and can be a dangerous practice. Placing the forearm in this position greatly limits and potentially eliminates the ability to lift the patient's lower jaw and tongue. In effect, it promotes the use of the teeth as a fulcrum as the only lifting option.

Step 10 Introduce the ET tube and advance it to the proper depth.

Upon visualization of the vocal cords, pass the tube through the cords and into the trachea.

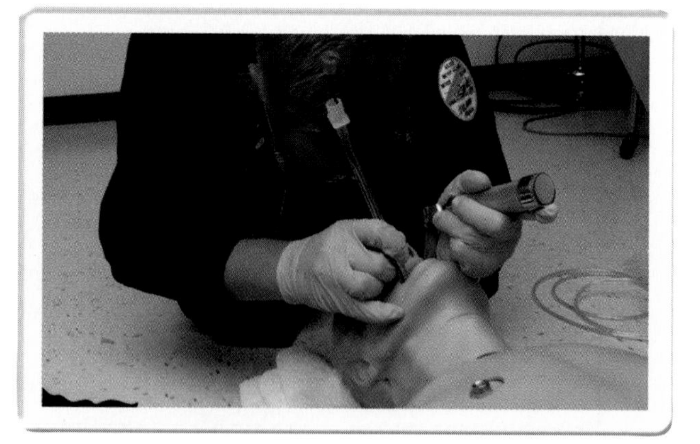

As the cuff of the tube passes the cords, stop the insertion process. Usually this will place the tube at a depth of 22 to 24 cm at the patient's teeth. This is roughly three times the diameter of the tube. Be sure to document the position of the tube relative to the teeth.

If after two attempts you are unable to perform endotracheal intubation, strongly consider using an alternative airway device or procedure.

Step 11 Remove the laryngoscope from the patient's mouth and remove the stylet from the ET tube.

Remove the laryngoscope from the patient's mouth.

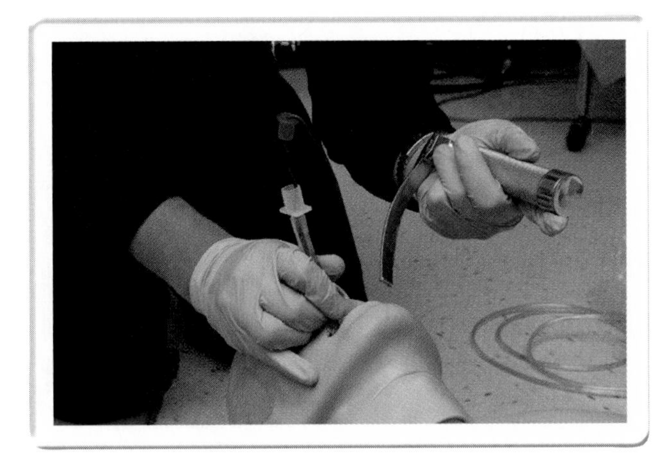

Remove the stylet from the ET tube.

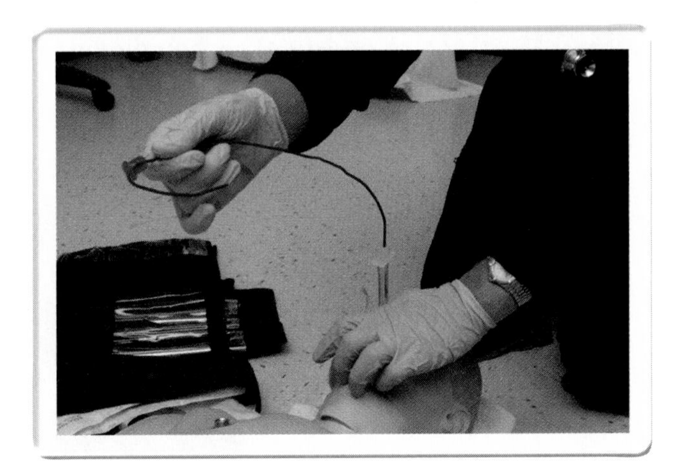

In the Field

Inflating the Endotracheal Cuff

The volume of air used in the ET cuff is based on the size of the tube and the size of the patient. Remember that the cuff's purpose is to seal the airway and that the airway size will be different in each patient. Conditions of edema and trauma will also alter the size of the airway. As the cuff is inflated, be sure to place only enough air to fill the balloon. Rarely will this be a full 10 mL of air. Overinflation of the cuff can lead to tracheal injury from excessive pressure acting as an internal tourniquet.

Step 12 Inflate the distal cuff to proper pressure and immediately disconnect syringe.

Inflate the distal cuff of the ET tube with 5 to 10 mL of air. Most patients will need less than the full 10 mL to fill the cuff. Remove the syringe from the inflation port.

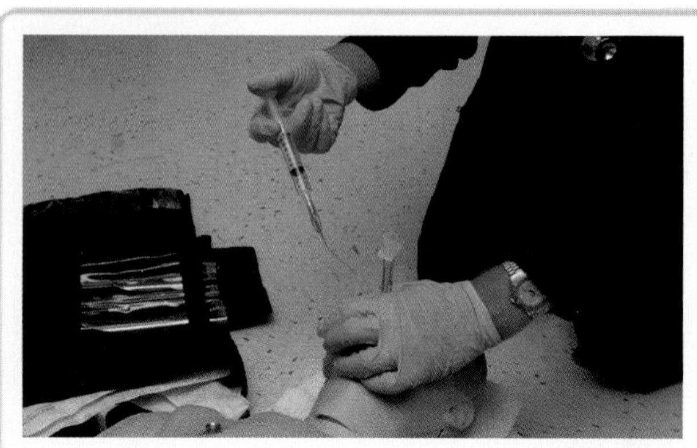

Step 13 Attach an end-tidal carbon dioxide detector.

Attach an end-tidal carbon dioxide detector to the ET tube.

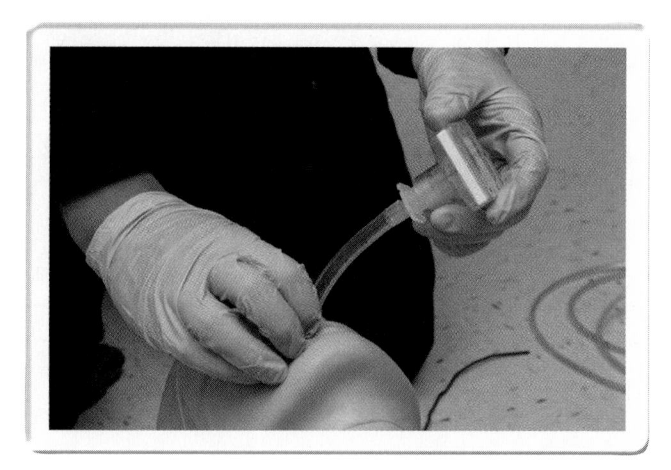

Step 14 Direct ventilation of the patient.

While still holding the tube in place, have your partner attach the bag-mask device to the endotracheal tube and begin ventilations.

15 Confirm proper placement by auscultation over epigastrium and bilaterally over each lung.

Auscultate the abdomen and the chest to determine correct tube placement. The preferred sequence is to listen over the gastric region first, and then over the lower left, upper left, upper right, and finally the lower right chest. By using this sequence you can determine incorrect placement with a minimal amount of unnecessary and useless ventilations. After confirming placement, move to the next step.

If air is heard over the epigastric region, remove the tube and begin ventilating the patient with a bag-mask device. Some practitioners prefer to leave ET tubes in the esophagus upon finding missed intubation, preferring instead to intubate around this first tube. Their philosophy is that this reduces the chances of missing a second time. However, this may confuse the landmarks and definitely limits the visibility in the hypopharynx. A problem of inefficient ventilations is also created.

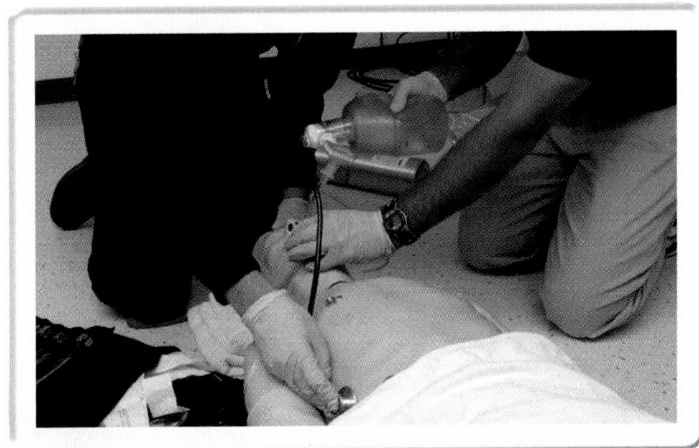

Absent breath sounds on the left are an indication of right mainstem intubation. Deflate the cuff. Gently and carefully pull back on the ET tube until breath sounds are heard over the left chest. Reinflate the cuff. Note and document the tube's position in relation to the teeth.

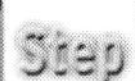

16 Secure the ET tube.

Maintaining a hold on the tube, attach a commercially made endotracheal tube holder (shown below). Commercial devices are preferred over other techniques because they provide the most effective control of the airway. These devices also eliminate the need for an oral airway, because they include a bite block.

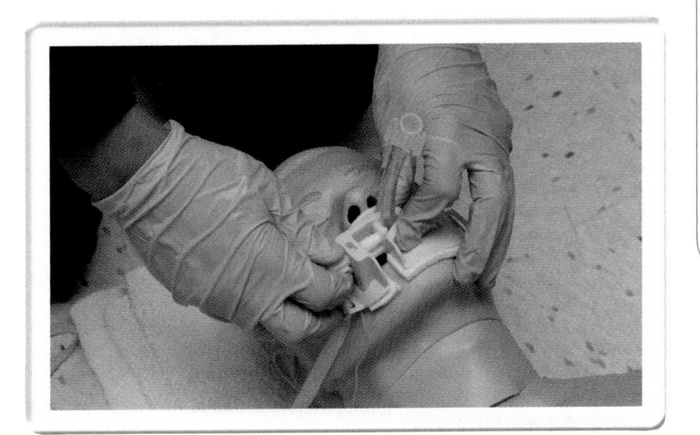

In the Field

Commercial Tube Holders

Because patients intubated in the prehospital setting will be lifted, pulled and pushed, transported at high speed, wheeled on a stretcher, and face the miscommunications of movement attempts, commercially made ET tube holders are recommended.

Step 16 continued

If a commercially made device is not available, secure the ET tube using tape.

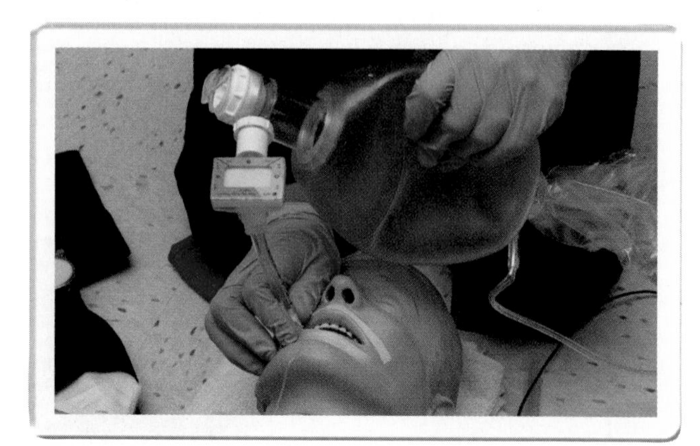

If you have secured the ET tube using tape, place a bite block in the patient's mouth.

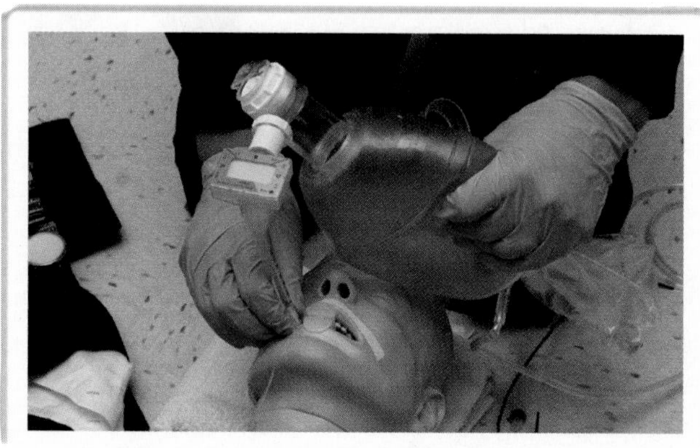

In the Field

Monitoring Tube Placement

By far the most effective tool in monitoring tube placement is the use of continuous capnographic monitoring. Consistent waveforms and carbon dioxide readings are evidence of proper tube placement as well as the effectiveness of ventilation and oxygenation. Even though the tube's placement has been ensured, continual reassessment should occur. This should be performed any time the patient is moved or defibrillated. Also, always ensure and document the tube placement before turning over patient care to the receiving facility or to a transport team.

Step 17 Use three means to ensure tube placement.

Reassess tube placement through assessment of gastric and breath sounds. At least two additional means of ensuring tube placement should be used. These include the following:

- Visualization of the tube passing between the cords
- End-tidal carbon dioxide detectors
- Fogging of the ET tube
- Esophageal detectors
- Capnographic monitors

Once proper placement is ensured, an appropriately sized extrication collar should be applied to maintain the position of the head and neck, and thus the airway.

All findings should be verbalized so that all members of the team are aware of the patient's airway status.

Pediatric Endotracheal Intubation

Performance Objective

Given a pediatric patient and appropriate equipment, the candidates shall insert an endotracheal (ET) tube and verify proper placement using the criteria herein prescribed, in 6 minutes or less.

Equipment

The following equipment is required to perform this skill:

- Appropriate body substance isolation/personal protective equipment
- Oxygen cylinder, regulator, and key
- Oxygen delivery device (appropriate to patient)
  - Bag-mask device
- Oropharyngeal airway (pediatric sizes)
- Laryngoscope handle and blades (straight and curved)
- Endotracheal tubes, uncuffed (2.5 to 5.5)
- Malleable intubation stylet (appropriate size for ET tube)
- Water-soluble lubricant
- Towel or other padding
- Colorimetric end-tidal carbon dioxide detector (infant/child size)
- Tube holder or tape
- Stethoscope
- Suction device and tubing
- Suction catheter
  - Rigid wand (Yankauer type)
  - Flexible catheter

Equipment that may be helpful:

- Pulse oximeter
- End-tidal carbon dioxide meter
- Spare bulb, batteries

Indications

- Ensure definitive airway
- Assist with long-term ventilation

Contraindications

- None

Complications

- Unrecognized and uncorrected esophageal intubation
- Right mainstem bronchial intubation
- Laryngeal injury
- Dental injury

Procedures

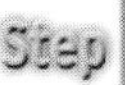 

Step 1 Ensure body substance isolation before beginning procedures.

Prior to beginning patient care, appropriate body substance isolation procedures should be employed.

Step 2 Open the airway manually.

Before opening the airway, consider the possibility of cervical spine injury. If spinal injury is suspected, use the jaw-thrust maneuver to open the airway. If no spinal cord injury is suspected, open the child's airway by performing the head tilt–chin lift maneuver.

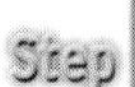 

Step 3 Elevate the child's tongue and appropriately insert simple adjunct.

Measuring from the corner of the mouth to the base of the earlobe, choose the correct-sized airway for the child.

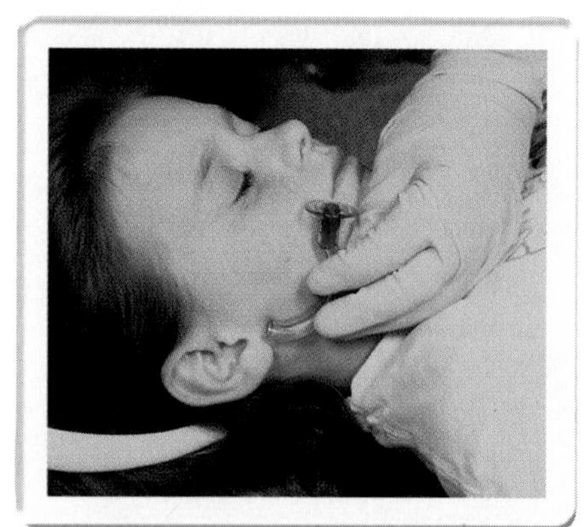

Gently open the airway and insert the oropharyngeal airway from the side of the mouth with the tip toward the inside of the cheek. A tongue depressor can be used to hold the tongue inferior and the airway inserted tip down.

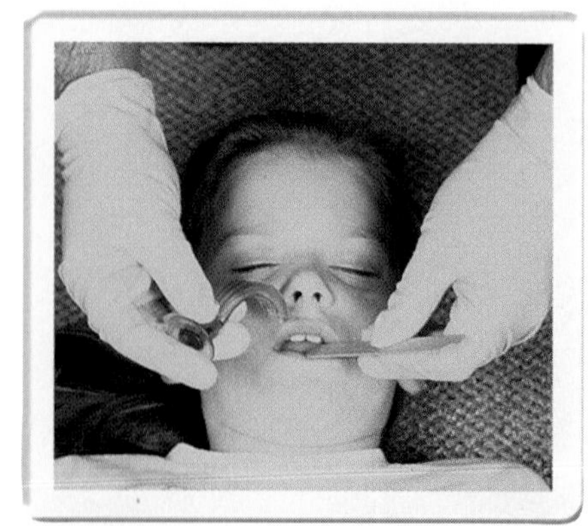

continued

If the child starts to gag, remove the airway by pulling the flange anterior and inferior. Be aware that a gag reflex in children can lead to serious bradycardia.

Ventilate the child immediately with bag-mask device.

Select the appropriate-sized bag for the child. Bag and mask sizes are normally listed as adult, child, and infant. These sizes are adequate for the average-sized person meeting the age definition criteria. However, large children will require a bag that will deliver the appropriate volume. Likewise, pediatric bags may be appropriate for small adults or adults in whom overpressurization can cause pulmonary damage. Masks should be chosen to fit the child without air leakage.

Position the mask with the apex over the bridge of the nose and the base between the lower lip and the prominence of the chin. Begin ventilations as soon as the mask is sealed and assess for air leakage.

The initial ventilations of a nonbreathing child should be performed in a controlled and accelerated manner. Deliver oxygen for at least 30 seconds. Use extreme caution not to overventilate. Each breath should be given slowly, with adequate time for exhalation in between.

After 30 seconds of preoxygenation, deliver ventilations at a rate of 10 to 12 breaths/min, with a tidal volume adequate to create chest rise.

Pay close attention to ensure that each ventilation is of the appropriate volume and of a consistent rate.

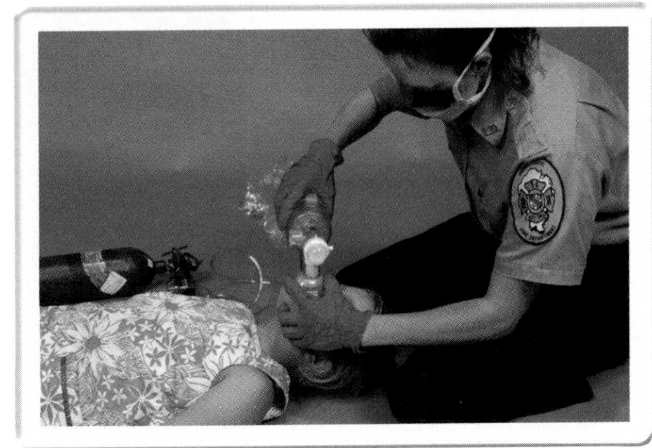

Allow adequate exhalation between each breath.

Effective ventilations are those breaths that cause the chest to rise adequately for the size of the patient. In addition to assessing chest rise, consider checking breath sounds. Be cautious not to overventilate. Each ventilation should be delivered slowly and easily, lasting 1 second each. Fast and high-pressure ventilations can cause air to enter the stomach, increasing the risk of vomiting. The Sellick maneuver should never be performed on children or infants. Although it will be necessary to stop ventilations for assessment and other procedures, you should never allow a nonbreathing patient to go without ventilations for longer than 30 seconds.

Step 5 Provide supplemental oxygen.

If not previously performed, attach oxygen tubing to the inlet on the bag-mask device and to the oxygen source. Open the cylinder valve and adjust liter flow to between 12 and 15 L/min (see Skill 4).

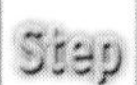 

Step 6 Direct assistant to take over ventilations.

Direct a qualified assistant to take over ventilations.

Step 7 Select, check, and prepare the equipment.

While the child is being hyperventilated, the person responsible for the intubation must ensure that all the necessary equipment is available and in working order. Ensure that the ET tube is the proper size for the child. Proper sizing can be achieved by comparing the diameter of the ET tube to the diameter of the child's little finger or the opening of the child's nares. Under most normal circumstances, an intubation stylet will not be necessary. If a stylet is used, make sure it is positioned so that the end of the stylet does not extend past the Murphy's eye.

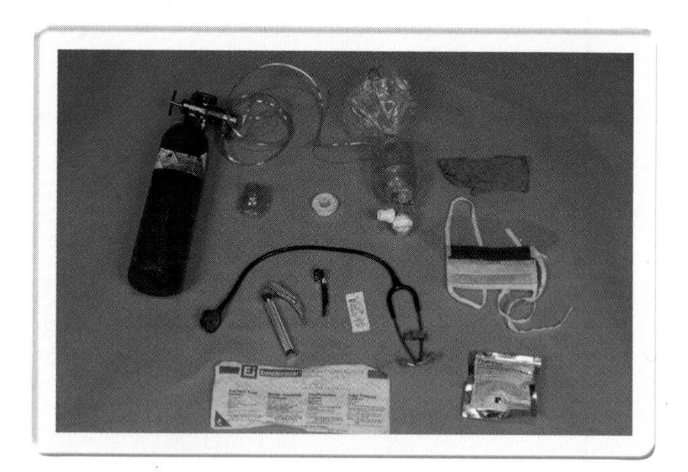

Uncuffed ET tubes should be used when intubating children younger than 8 years. In the event you encounter an ET tube with a cuff in smaller sizes (designed for specialty and veterinary purposes), resist the temptation to inflate the cuff by cutting the inflation port from the tube. In cases where the chosen tube is too small for the child, slight inflation of the cuff to seal the airway may be necessary. Be sure that this is the exception and not the rule, and monitor the amount and pressure of the air in the cuff.

Assemble the laryngoscope with the desired blade. For smaller infants and neonates a No. 1 Miller is the preferred blade (Macintosh blades are contraindicated because of anatomic differences in a child's upper airway). Ensure that the bulb is tight in the blade. Check the brightness of the bulb. A bright white light is best for

Step 7 continued

all intubations. A yellow or dull light will cause the tissues of the hypopharynx to take on similar color schemes and result in difficulties visualizing landmarks and the vocal cords. Make sure the light is turned off until the laryngoscope is placed in the child's mouth.

Choosing the Right-Sized Tube and Blade

Age	Premature	Newborn	6 months	1-2 years	4-6 years	8-12 years
Tube size	2.5	3-3.5	3.5-4	4-5	5-5.5	6-7
Blade size	0	0-1	1	1-2	2	2-3

Finally, always ensure that suction is available and working before beginning intubation. Both a rigid suction tip and a flexible catheter, appropriate for the size of the endotracheal tube, should be available.

Step 8 Place child in neutral or sniffing position, placing pad under the child's torso.

Gently extend the child's neck so that the head is positioned with the facial plane approximately 25° to 30° from the floor (just slightly extended from the neutral position). Place a 1″ pad beneath the shoulders of the child.

Step 9 Insert blade while displacing the tongue.

Carefully insert the blade of the laryngoscope into the child's mouth to the depth of the uvula. Using a sweeping motion, move the child's tongue to the left as you lift at an angle along the facial plane, keeping the laryngoscope off the teeth.

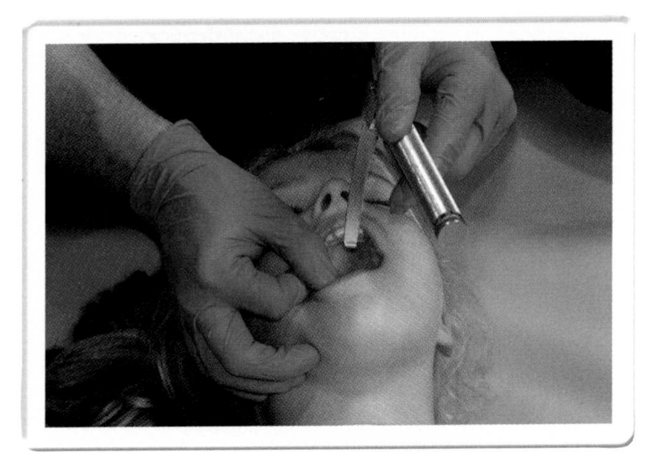

Step 10 Introduce the ET tube and advance it to the proper depth.

Upon visualization of the vocal cords, pass the tube through the cords and into the trachea.

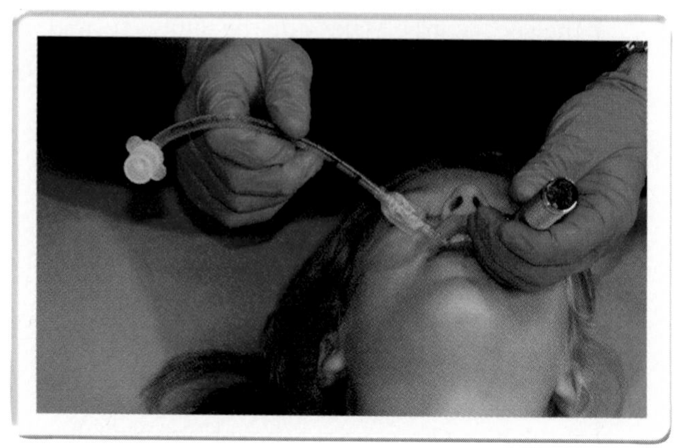

As the eye of the tube passes the cords, stop the insertion process. Many smaller-diameter tubes have a black line that indicates the position of the vocal cords. If after three attempts you are not able to achieve endotracheal intubation, consider allowing another practitioner to perform the procedure, or consider immediate transport without a secure airway.

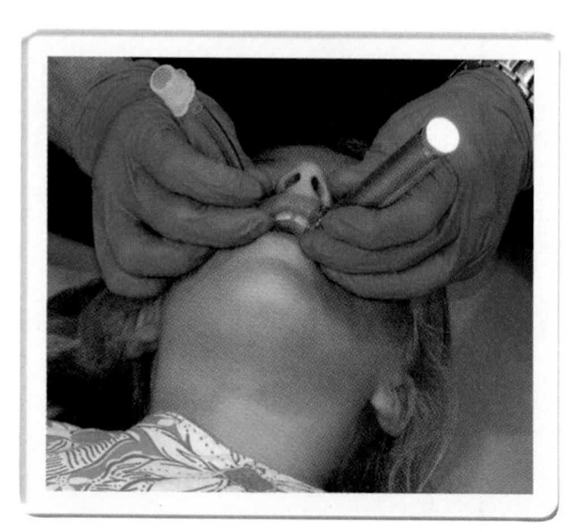

Step 11 Remove the laryngoscope from the child's mouth.

Remove the laryngoscope from the child's mouth.

Step 12 Attach an end-tidal carbon dioxide detector.

Attach an end-tidal carbon dioxide detector to the ET tube.

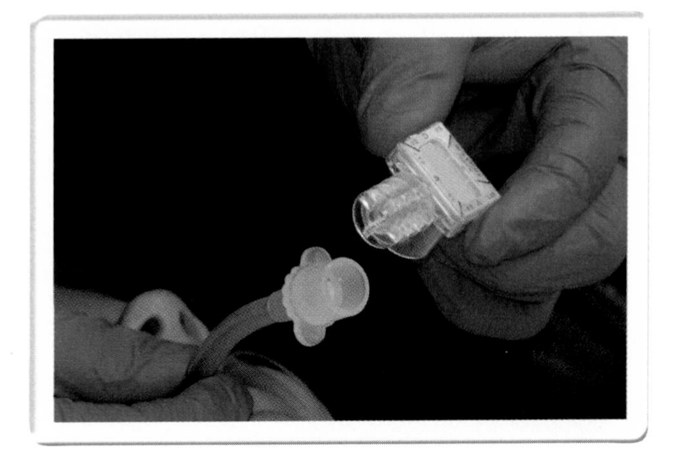

Step 13 Direct ventilation of the child.

While still holding the tube in place, have your partner attach the bag-mask device to the endotracheal tube and begin ventilations.

Instruct your partner to deliver breaths as you move your stethoscope over the child's abdomen and chest.

Step 14 Confirm proper placement by auscultation over epigastrium and bilaterally over each lung.

Auscultate the abdomen and the chest to determine correct tube placement. The preferred sequence is to listen over the gastric region first, and then over the lower left, upper left, upper right, and finally the lower right chest. By using this sequence, you can determine incorrect placement with a minimal number of unnecessary ventilations.

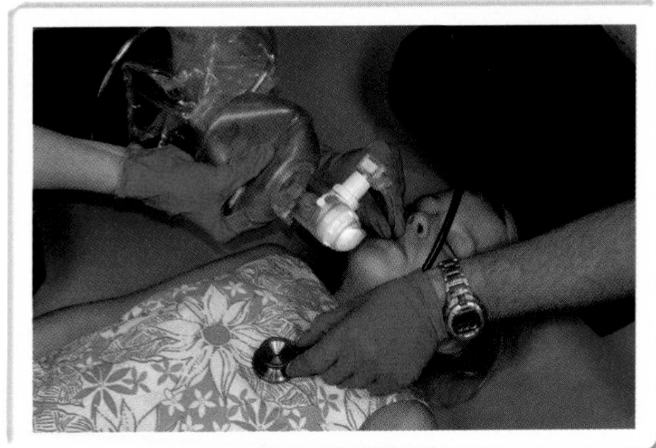

Step 14 continued

In infants and small children, return the patient's head and neck to a neutral position and reassess the breath sounds. Be aware that any extension of the neck can reposition the tube into a right mainstem intubation, while a flexion of the neck can pull the tube out of the larynx.

If air is heard over the epigastric region, remove the tube and begin ventilating the patient with a bag-mask device. Some practitioners prefer to leave endotracheal tubes in the esophagus upon finding a missed intubation, preferring instead to intubate around this first tube. *This should never be performed in a pediatric patient.*

Absent breath sounds on the left are an indication of right mainstem intubation. Gently and carefully pull back on the ET tube until breath sounds are heard over the left chest.

Safety Tips

Neonate and Infant Intubation

In neonates and smaller infants, right mainstem intubations can cause pneumothoraces to develop. Be alert for a tension pneumothorax to develop and inform the receiving physician of the initial right mainstem intubation.

Step 15 Secure ET tube.

Note the placement of the distance marker at the child's teeth or gums. Maintaining a hold on the tube, encircle the tube with tape at the level of the child's gums.

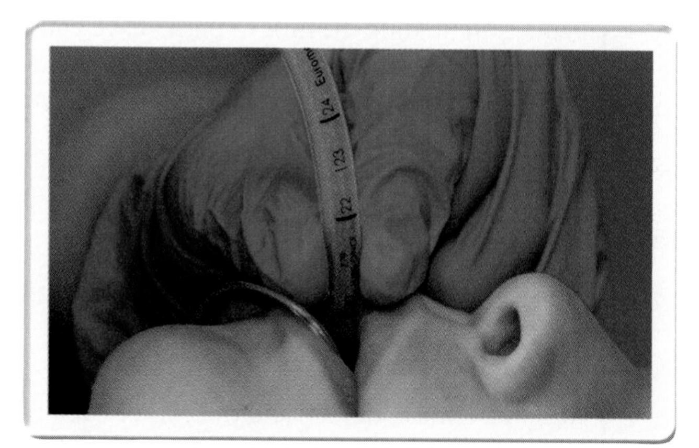

Wrap the tape around the child's neck and again encircle the tape around the tube at the level of the child's gums. Ensure that you have not placed undue pressure on the vessels of the neck by placing the tape over the prominence of the anterior mandible.

Step 16 Use three means to ensure tube placement.

Reassess tube placement through assessment of gastric and breath sounds. At least two additional means of ensuring tube placement should be used. These include the following:

- Visualization of the tube passing between the cords
- End-tidal carbon dioxide detectors (pediatric size)
- Fogging of the ET tube
- Capnographic monitors

Finally, once placement is ensured, an appropriately sized extrication collar should be applied to maintain the position of the head and neck, and thus the airway.

All findings should be verbalized so that all members of the team are aware of the patient's airway status.

Even though the tube's placement has been ensured, continual reassessment should occur. This should be performed any time the child is moved or defibrillated. Also, always ensure and document the tube placement before turning the child over to the receiving facility or to a transport team.

SKILL 49

Tracheal Suctioning

Performance Objective

Given a patient with an endotracheal (ET) tube in place and appropriate equipment, the candidate shall suction down the ET tube, using sterile procedures, in 3 minutes or less.

Equipment

The following equipment is required to perform this skill:

- Appropriate body substance isolation/personal protective equipment
- Oxygen cylinder, regulator, and key
- Oxygen delivery device (appropriate to patient)
  - Bag-mask device
- Suction device
- Suction tubing
- Suction catheter
  - Rigid wand (Yankauer type)
  - Flexible catheter
- Sterile water

Indications

- Secretions in the endotracheal tube
- Removal of mucus plugs in patients with asthma

Contraindications

- Hypoxemia
- Atelectasis
- Cardiac arrhythmias
- Tracheobronchial trauma
- Increased intracranial pressure (from excessive cough reflex and agitation)

Complications

- Hypoxemia
- Atelectasis

Procedures

Step 1 Ensure body substance isolation before beginning procedures.

Gloves and eye protection are essential. Masks are highly recommended if the patient has any possibility of a respiratory disease. Because this is an aseptic procedure, sterile gloves will be worn. It is recommended that the sterile gloves be donned over nonsterile gloves for greater protection of the practitioner.

Step 2 Ensure sterility throughout the procedure.

The suction catheter placed into the airway should be sterile when inserted. Maintenance of the sterile field through the use of sterile gloves and procedures is essential. Should the sterile field be contaminated, a fresh sterile field should be created to prevent respiratory infections.

Step 3 Identify and select appropriate suction catheter.

Endotracheal suctioning is performed using a flexible suction catheter. The size of the catheter is determined by the size of the internal diameter of the endotracheal tube. The catheter should be small enough to pass the curves of the endotracheal tube, and large enough to provide adequate suction. A catheter that is too small will have difficulty suctioning the entire airway and may not pick up thick secretions. This will reduce the efficiency of each suctioning attempt, prompting the practitioner to make repeated attempts to clear the airway and thus increasing the possibility of complications and injury.

ET Tube Size (Inside Diameter)	Catheter Size
8.0 mm	12-16F
7.0 mm	10-12F
6.0 mm	10F
5.0 mm	10F
4.0 mm	8-10F
3.0 mm	6F

Step 4 Prepare equipment.

Prepare the suctioning system to provide a negative pressure of 100 to 150 mm Hg. Using an aseptic technique, open the catheter package. Don the sterile glove on the dominant hand. Many suction packages come with a collapsed water cup. If this is present, open the cup with the sterile hand and have an assistant fill the cup with sterile water. If no assistant is available, you may fill the cup with sterile water using the nondominant hand. If the suction package does not have a water cup, open a bottle of sterile water and have it ready to clean the catheter as necessary.

Step 5 Preoxygenate the patient.

Provide three to five quick breaths to hyperoxygenate the patient.

Step 6 Mark maximum insertion length with thumb and forefinger.

The catheter should be long enough to extend approximately 2″ past the end of the ET tube in the adult patient, and no more than 1″ in pediatric patients. Estimate this distance and mark the location with the thumb and forefinger of your dominant hand. This will be your maximum insertion depth.

Most practitioners prefer to insert the catheter until resistance is felt. This is an accepted practice and in many ways preferred.

Insert catheter into the ET tube, leaving catheter port open.

Using the nondominant hand, remove the bag-mask device from the ET tube. Carefully, observing aseptic procedures, insert the catheter into the endotracheal tube. Continue to insert the catheter until resistance is felt or you have reached your predetermined maximum depth. For thick secretions, instillation of 3 to 5 mL of sterile water down the ET tube may be helpful.

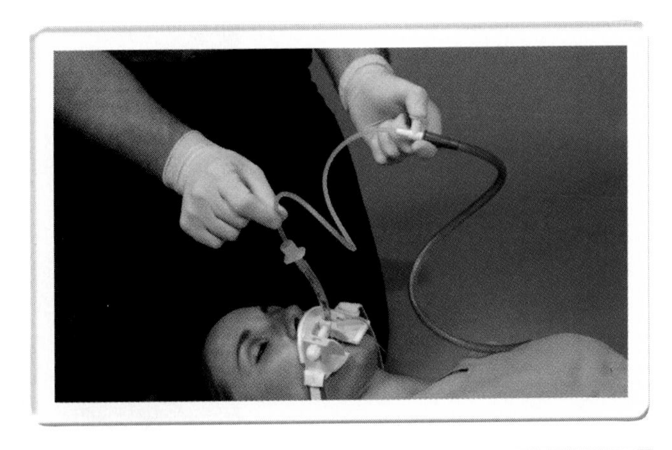

At proper insertion depth, cover the catheter port and apply suction while withdrawing the catheter.

Pull the catheter back approximately 1 cm from the point where resistance is felt and apply suction by covering the catheter port with the thumb of the nondominant hand. Carefully suction while removing the catheter. A twisting motion that includes slight advancements back into the tube should be used. Limit suctioning to no more than 10 seconds. Use the dominant hand to guide the sterile portion of the catheter out of the tube. Keep it away from nonsterile surfaces. Wrapping the catheter in the dominant hand will help protect the sterile environment.

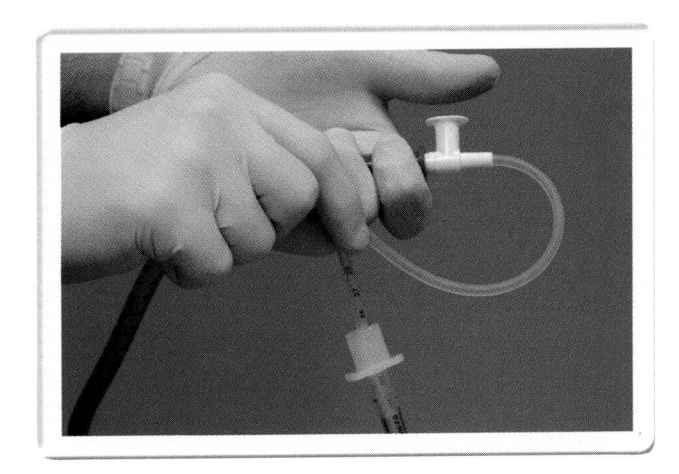

Step 9 Ventilate or direct the ventilation of the patient as the catheter is flushed with sterile water.

Using the nondominant hand, reattach the bag-mask device and return to ventilation. Using the dominant hand, insert the suction catheter into the prepared sterile water and clean the tip by applying short periods of suction.

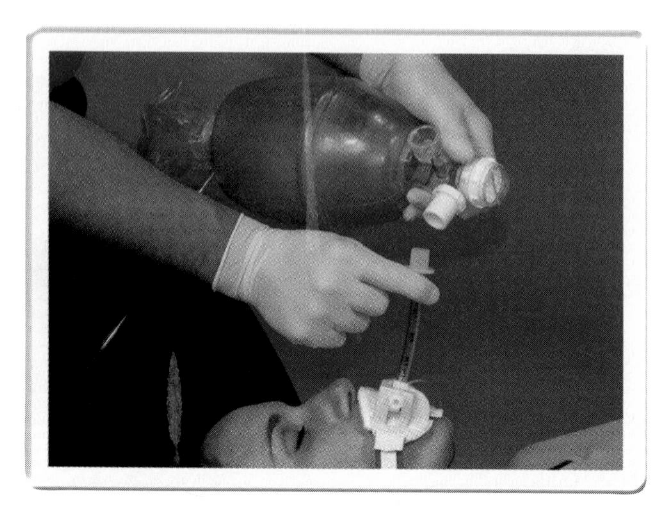

Step 10 Repeat suction as necessary.

Following the same procedures, continue to suction until the airway is clear.

Check Your Knowledge

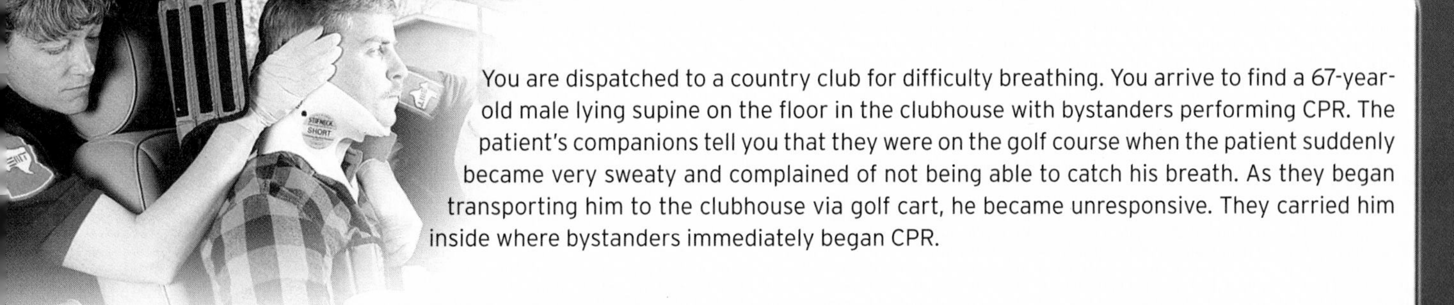

You are dispatched to a country club for difficulty breathing. You arrive to find a 67-year-old male lying supine on the floor in the clubhouse with bystanders performing CPR. The patient's companions tell you that they were on the golf course when the patient suddenly became very sweaty and complained of not being able to catch his breath. As they began transporting him to the clubhouse via golf cart, he became unresponsive. They carried him inside where bystanders immediately began CPR.

1. After manually opening the patient's airway, what is the next step you should take?

A. Combitube placement
B. KING LT-D placement
C. Endotracheal intubation
D. Insert simple airway adjunct

2. What is the longest period of time you should allow a nonbreathing patient to go without ventilations while you perform interventions?

A. 10 seconds
B. 15 seconds
C. 30 seconds
D. 1 minute

3. When performing endotracheal intubation for this patient, when should you stop advancement of the ET tube?

A. When you encounter resistance
B. As the cuff of the tube passes the cords
C. When the patient gags
D. At a depth of 20 cm

4. What volume of air should you use to inflate the distal cuff of the ET tube?

A. 0-5 mL
B. 5-10 mL
C. 10-15 mL
D. 15-20 mL

5. After assessment of gastric and breath sounds, which of the following methods may be used to further ensure proper tube placement?

A. End-tidal carbon dioxide detector
B. Esophageal detector
C. Capnographic monitor
D. All of the above

Additional Questions

6. Which of the following is contraindicated when intubating a small infant or neonate?

A. Macintosh blade
B. Miller blade
C. Simple airway adjuncts
D. Supplemental oxygen

7. What is the prescribed amount of air for inflation of the esophageal cuff in an adult Combitube?

A. 5 mL
B. 15 mL
C. 85 mL
D. 100 mL

8. Proper placement of a Combitube can be confirmed by which of the following methods?

A. Observing chest rise
B. Auscultation over the epigastrium
C. Auscultation bilaterally over each lung
D. All of the above

9. Which size suction catheter should be used with a 7.0 mm ET tube?

A. 6F
B. 8-10F
C. 10-12F
D. 12-16F

10. When performing tracheal suctioning, the suction catheter should extend approximately 2" past the end of the ET tube in an adult patient, and no more than ______ in a pediatric patient.

A. 1/2"
B. 1"
C. 1 1/2"
D. 2"